Boundary Elements
and
Other Mesh Reduction Methods
XXX

WITPRESS

WIT Press publishes leading books in Science and Technology.
Visit our website for the current list of titles.
www.witpress.com

WITeLibrary

Home of the Transactions of the Wessex Institute.
Papers presented at BEM/MRM XXX are archived in the WIT elibrary in volume 47 of WIT Transactions on Modelling and Simulation (ISSN 1743-355X).
The WIT electronic-library provides the international scientific community with immediate and permanent access to individual papers presented at WIT conferences.
http://library.witpress.com

Thirtieth World Conference on Boundary Elements and Other Mesh Reduction Methods

BEM/MRM XXX

Conference Chairmen

L. Škerget
University of Maribor, Slovenia

C.A. Brebbia
Wessex Institute of Technology, UK

International Scientific Advisory Committee

Organised by
Wessex Institute of Technology, UK

Sponsored by
International Journal of Engineering Analysis with Boundary Elements (EABE)

WIT Transactions

Transactions Editor

Carlos Brebbia
Wessex Institute of Technology
Ashurst Lodge, Ashurst
Southampton SO40 7AA, UK
Email: carlos@wessex.ac.uk

Editorial Board

Boundary Elements
and
Other Mesh Reduction Methods
XXX

Editors

L. Škerget
University of Maribor, Slovenia

C.A. Brebbia
Wessex Institute of Technology, UK

WIT*PRESS* **Southampton, Boston**

Editors:

L. Škerget
University of Maribor, Slovenia

C.A. Brebbia
Wessex Institute of Technology, UK

Published by

WIT Press
Ashurst Lodge, Ashurst, Southampton, SO40 7AA, UK
Tel: 44 (0) 238 029 3223; Fax: 44 (0) 238 029 2853
E-Mail: witpress@witpress.com
http://www.witpress.com

For USA, Canada and Mexico

Computational Mechanics Inc
25 Bridge Street, Billerica, MA 01821, USA
Tel: 978 667 5841; Fax: 978 667 7582
E-Mail: infousa@witpress.com
http://www.witpress.com

British Library Cataloguing-in-Publication Data

A Catalogue record for this book is available
from the British Library

ISBN: 978-1-84564-121-4
ISSN: (print) 1746-4064
ISSN: (on-line) 1743-355X

The texts of the papers in this volume were set individually by the authors or under their supervision. Only minor corrections to the text may have been carried out by the publisher.

Printed in Great Britain by Cambridge Printing

Preface

We have been very proud at the University of Maribor that the 30th International Conference on Boundary Elements and other Mesh Reduction Methods was hosted in our town. Slovenia has just completed a term of the European Union presidency. For six months we were the nexus for four hundred million plus united Europeans. We are not accustomed to such attention. We were similarly very nervous when the organisers of the Conference chose us as the venue for the finale of the third decade of Boundary Elements, or rather, the beginning of the fourth.

Thirty years in the life of a new idea is, as in the case of a human being, a pivotal point; still young, but old enough; considered bold, yet wise; looking forward, and having something to show. Since 1978, we have come a long way and now BEM is a recognized and powerful computational method.

Our group, at the School of Engineering, continues to work in transport phenomena, predominantly in fluids. Our friends and colleagues specializing in solids can take advantage of parallelization and problem transfer to the boundary much better than we, who are concerned with fluid dynamic problems.

For thirty years, we have been spared the problems that other methods in the area of fluids could not avoid – i.e. the behaviour of pressure on account of vorticity–velocity type formulations. On the other hand, we were concerned with questions others knew nothing or little about – full system matrices, discussing memory sharing and file storage and other topics more suitable for computer scientists than engineers.

We can solve more complex problems than we ever dreamed about. The number of nodes is now in the order of thousands, rather than hundreds. But the quest continues and the goal is still far off as some questions remain unanswered. In our field, problems such as turbulence are still difficult to solve, as not all problems can be transferred to the boundary. Compromise was necessary in the past and was achieved by using boundary domain techniques. Computers have become faster and cheaper, but we have learnt that 'brute' force is not always the best approach. Sometimes it pays to step back and think about the problem to be solved.

It is time to pass these challenges on to a new generation to continue further developing the beauty and elegance of the boundary element and other mesh reduction methods.

Leopold Škerget
University of Maribor, 2008

Carlos Brebbia writes:

Leopold Škerget's contribution to the development of Boundary Elements applications in fluid dynamics has been outstanding and of major international significance. The result of this work has been the creation of a research school at the University of Maribor, second to none in the field, leading to a new generation of researchers involved in solving a range of ever larger and more complex problems.

The School at Maribor has pioneered work on BEM in fluid dynamics all over the world and they shared their results in the spirit of generosity which is the hallmark of true scientists.

The Boundary Element community owes a debt of gratitude to Leopold's group and this is the main reason that the 30th Conference was held at the University of Maribor. On behalf of all of us at the Conference, I am also grateful to all contributors for participating in this special meeting, and particularly to those who helped to select the papers published in this book.

Carlos A. Brebbia
Maribor, 2008

Contents

Section 3: Electrical engineering and electromagnetics

Section 4: Damage mechanics and fracture

Section 5: Mesh reduction techniques

Section 6: Advanced computational techniques

Section 1
Fluid flow

Towards a fast single domain – subdomain BEM algorithm for 3D incompressible fluid flow

J. Ravnik, L. Škerget & Z. Žunič
University of Maribor, Faculty of Mechanical Engineering,
Smetanova ulica 17, SI-2000 Maribor, Slovenia

Abstract

Work towards acceleration and computer memory reduction of an algorithm for the simulation of laminar viscous flows by solving of 3D velocity-vorticity formulation of the Navier–Stokes equations is presented. The algorithm employs a combination of a subdomain boundary element method (BEM) and fast single domain BEM. The single domain BEM on a Poisson type equation is employed to calculate boundary vorticity values. After discretization, the single domain BEM algorithm yields a fully populated system of linear equations. The non-homogenous part of the Poisson equation yields a fully populated matrix of domain integrals. In order to lower the computational demand, a fast multipole expansion algorithm is used on the domain matrix. The fundamental solution is expanded in terms of spherical harmonics. The computational domain and its boundary are recursively cut up forming a tree of clusters of boundary elements and domain cells. Compression is achieved in parts of the matrix, which correspond to boundary-domain cluster pairs that are admissible for expansion. Significant reduction of the complexity is achieved. The paper presents results of initial testing of the FMM algorithm.
Keywords: fast multipole method, boundary element method, Poisson equation.

1 Introduction

Our research group works on the application of the Boundary Element Method to fluid flow problems. Recently Ravnik *et al.* [1] developed a 3D subdomain - single domain BEM numerical scheme for solving incompressible velocity-vorticity formulation of Navier–Stokes equations. A crucial part of this algorithm is the calculation of boundary vorticity values. They are obtained by solving a Poisson type partial differential equation using single domain BEM. The domain integral, which arises from the non-homogenous part of the Poisson equation, requires the

WIT Transactions on Modelling and Simulation, Vol 47, © 2008 WIT Press
www.witpress.com, ISSN 1743-355X (on-line)
doi:10.2495/BE080011

discretization of the domain and the calculation of a large number of domain matrix elements, thus limiting the maximal mesh size. In this work, we are presenting the application of the Fast Multipole Method (FMM) on the Poisson type equation for a data sparse representation of the non-homogenous part.

The origins of FMM can be found in a fast algorithm for particle simulations developed by Greengard and Rokhlin [2]. The algorithm decreases the complexity of interacting particles from (n^2) to linear complexity. The method was since then used by many authors for a wide variety of problems using different expansion strategies. Recently Bui *et al.* [3] combined FMM with the Fourier transform to study multiple bubbles dynamics. Gumerov and Duraiswami [4] applied the FMM for the biharmonic equation in three dimensions. The boundary integral Laplace equation was accelerated with FMM by Popov and Power [5] and Popov *et al.* [6].

2 BEM for Poisson equation

The Poisson equation is a partial differential equation including a diffusion operator and a non-homogenous right-hand side,

$$\nabla^2 u(\vec{r}) = b(\vec{r}); \quad \vec{r} \in \Omega, \tag{1}$$

where the unknown scalar field function $u(\vec{r})$ and the non-homogenous source term $b(\vec{r})$ are defined in a domain Ω. The solution of such problems can be found when suitable boundary conditions are applied, i.e. known scalar function or its flux ($q = \vec{n} \cdot \vec{\nabla} u$) on the boundary $\Gamma = \partial\Omega$. An integral form of Poisson type equation for a scalar field function $u(\vec{r}) \in \Omega$ is (Wrobel [7]):

$$c(\vec{\xi})u(\vec{\xi}) + \int_\Gamma u(\vec{r})\vec{n} \cdot \vec{\nabla} u^\star d\Gamma = \int_\Gamma q(\vec{r}) u^\star d\Gamma - \int_\Omega b(\vec{r}) u^\star d\Omega; \qquad \vec{\xi} \in \Gamma, \tag{2}$$

where $\vec{\xi}$ is the collocation point on the boundary, $\vec{n}$ is the unit normal and $u^\star = 1/4\pi|\vec{r} - \vec{\xi}|$ is the fundamental solution of the Laplace equation in 3D. The domain is approximated by domain cells $\Omega \approx \sum_{c=1}^{n_c} \Omega_c$ and its boundary by boundary elements $\Gamma \approx \sum_{e=1}^{n_e} \Gamma_e$. Within each domain cell and boundary element the field functions are approximated using domain Φ, boundary φ and boundary flux ϕ shape functions. In this paper domain cells are hexahedra and boundary elements are parallelepipeds. In each domain cell 27 nodes are used to achieve quadratic interpolation of function. Nine continuous nodes are used in boundary elements for quadratic interpolation of function and four discontinuous nodes are used to interpolate fluxes linearly. Considering these approximations in equation (2) we have:

$$c(\vec{\xi})u(\vec{\xi}) + \sum_{e=1}^{n_e} \sum_{i=1}^{9} u_i^e \int_{\Gamma_e} \varphi_i^e \vec{n} \cdot \vec{\nabla} u^\star d\Gamma$$

$$= \sum_{e=1}^{n_e} \sum_{i=1}^{4} q_i^e \int_{\Gamma_e} \phi_i^e u^\star d\Gamma - \sum_{c=1}^{n_c} \sum_{i=1}^{27} b_i^c \int_{\Omega_c} \Phi_i^c u^\star d\Omega. \tag{3}$$

The integrals are traditionally named as

$$h_i^{e,\vec{\xi}} = \int_{\Gamma_e} \varphi_i^e \vec{n} \cdot \vec{\nabla} u^\star d\Gamma, \quad g_i^{e,\vec{\xi}} = \int_{\Gamma_e} \phi_i^e u^\star d\Gamma, \quad \beta_i^{c,\vec{\xi}} = \int_{\Omega_c} \Phi_i^c u^\star d\Omega. \quad (4)$$

For a given collocation point $\vec{\xi}$ we must calculate integrals for each internal cell c, each boundary element e and all of the shape functions i. When the collocation point $\vec{\xi}$ is set into all boundary nodes, integrals may be arranged into matrices and the system of linear equations may be written in matrix-vector form. Let [ ] denote matrices and { } denote vectors. In matrix vector form equation (3) is:

$$[H]\{u\} = [G]\{q\} - [B]\{b\}. \quad (5)$$

Since all integral kernels are non zero, the matrices of equation (5) are fully populated. Considering the boundary conditions, the system of equations (5) is rearranged so that the unknown values of function and flux are gathered on the left side. A direct solver with LU decomposition is used to solve the resulting system. In order to evaluate the right-hand side of the system, we must calculate the domain matrix times vector product $[B]\{b\}$. Since the domain matrix is fully populated, we have lost the advantage of the boundary element method. The non-homogenous part of the Poisson equation requires the discretization of the domain and the calculation of a fully populated domain matrix.

The size of the boundary matrices $[H]$ and $[G]$ scale as number of boundary nodes squared. This are small compared to the domain matrix $[B]$, which scales as the number of boundary nodes times the number of domain nodes. Considering a cube with N nodes per edge, we can estimate the number of domain nodes as $n_d = N^3$, and the number of boundary nodes as $n_b \approx N^2$, thus the complexity of the domain integral matrix is $\mathcal{O}(N^5)$. Since, clearly, the domain contribution takes up most of the CPU time and storage space, this paper presents an application fast multipole method to obtain a data sparse representation of the domain matrix in such a way that the number of non-zero elements scales linearly with the number of nodes $\mathcal{O}(n_d)$.

3 Fast multipole method for the domain matrix

Let us consider the domain integral in equation (4). Since for each collocation point $\vec{\xi}$ integrals for all domain cells c must be evaluated, we are obviously faced by a problem of quadratic complexity. This is analogous to the problem of interaction of n particles (Barnes and Hut [8]), where the origins of the FMM can be found.

The method is based on the fact that it is possible to expand integral kernel, i.e. the fundamental solution, into a series and by doing so, separate the variables – the collocation point $\vec{\xi}$ and the domain integration point $\vec{r}$. In this work we will use spherical harmonics to expand the integral kernel into a series. Other expansions are also possible, such as Taylor series, Lagrangian polynomials, etc. In polar coordinate system $\vec{r} = (r, \varphi_r, \theta_r)$ and $\vec{\xi} = (\xi, \varphi_\xi, \theta_\xi)$ the integral kernel may be expanded into such series

$$\frac{1}{4\pi|\vec{r}-\vec{\xi}|} = \sum_{l=0}^{\infty}\sum_{m=-l}^{l} \frac{(-1)^m}{2l+1} \underbrace{\frac{1}{\xi^{l+1}} Y_l^{-m}(\theta_\xi, \varphi_\xi)}_{f(\vec{\xi})} \underbrace{r^l Y_l^m(\theta_r, \varphi_r)}_{g(\vec{r})}, \tag{6}$$

where the dependence on the collocation point and domain point are separate. The spherical harmonics may be calculated using a relationship to associated Legendre polynomials P_l^m:

$$Y_l^m(\theta, \varphi) = \sqrt{\frac{2l+1}{4\pi}\frac{(l-m)!}{(l+m)!}} P_l^m(\cos\theta) e^{im\varphi}. \tag{7}$$

The associated Legendre polynomials are evaluated using recurrence relations as described in Press *et al.* [9]. The domain integral of equation (4) may now be written as

$$\beta_i^{c,\vec{\xi}} = \sum_{l=0}^{\infty}\sum_{m=-l}^{l} F_l^m(\vec{\xi}) \int_{\Omega_c} G_l^{m,i}(\vec{r}) d\Omega. \tag{8}$$

We are now able to calculate each entry in the domain matrix with the above sum. The advantage of this becomes obvious when we consider a cluster of n_r nearby collocation nodes and a cluster of n_c nearby domain cells. These correspond to a $n_r \times n_c$ matrix block. Since the variables are separated, it is possible to evaluate two smaller matrix blocks ($n_r \times n_{\mathrm{exp}}$) and ($n_{\mathrm{exp}} \times n_c$), where n_{exp} is the number of expansion terms. Multiplication of the two smaller matrix block gives the full $n_r \times n_c$ matrix block up to an expansion error, which is defined by the number of terms in the expansion. In order for this technique to yield a data sparse representation of the full matrix block, the number of terms in the two smaller matrices should be smaller than the number of terms in the full matrix block, i.e.

$$2(n_r n_{\mathrm{exp}} + n_c n_{\mathrm{exp}}) < n_r n_c; \tag{9}$$

the factor 2 on the left-hand side is due to the fact that spherical harmonics are complex and must be stored as such, while real values are stored in the full matrix. As long as the collocation node cluster and the domain cells cluster are far apart from each other, we can expect a low number of expansion terms to yield a suitable approximation. When the clusters are nearby, they should be smaller and a larger number of expansion terms must be used. When the clusters coincide, i.e. the collocation nodes are a part of the integration cells, the kernel is singular. Such cluster pairs are called inadmissible and the corresponding matrix block is evaluated in full, without compression.

The collocation points on the boundary must be divided into clusters and coupled with clusters of domain cells. We constructed a tree of clusters of collocation points and a tree of clusters of domain cells. The trees were constructed in a recursive hierarchical manner. The problem domain is enclosed by a parallelepiped. All collocation points and all of the domain cells are within this root parallelepiped.

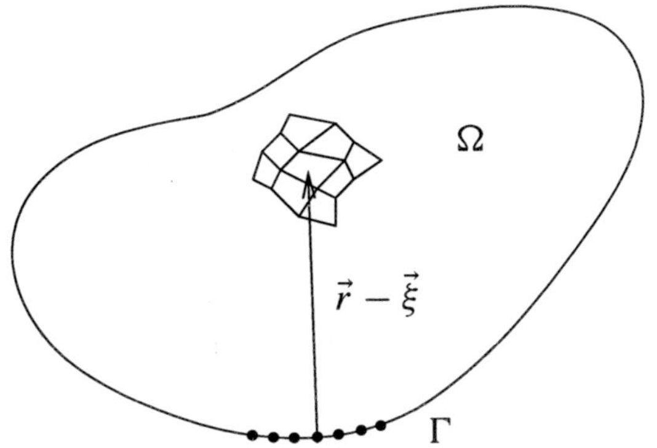

Figure 1: A problem domain shown with a cluster of collocation points $\vec{\xi}$ and a cluster of domain cells.

They make up root clusters of both trees. The parallelepiped is cut in half by a plane splitting the root clusters into two. The cutting process is repeated recursively, so the clusters on each level have less and less collocation points and domain cells.

With both cluster trees is place, the next step is to pair them, so a list of admissible and inadmissible matrix blocks can be formed. The admissible and inadmissible matrix blocks correspond to leaves on the tree of pairs of clusters. Each branch of the collocation tree is paired with each branch of the domain cells tree on the same level and with each branch of the domain cells tree on the next level forming branches on the tree of pairs of clusters. For each pair a decision is taken weather it is possible to do compression (i.e. the cluster pair is admissible) or lower level pairs must be considered. If admissibility is not reached until the last level, such pairs are inadmissible and will be calculated in full without compression.

The admissibility criterion is devised as follows. We are considering a cluster of collocation points and a cluster of domain cells. Firstly we try to find a origin of the coordinate system in nodes corresponding to domain cells in the cluster. We choose such origin that the ratio r/ξ is minimal for all pairs of collocation nodes and domain cells. If the minimal ratio is above one, series expansion is not possible for this pair of clusters, thus this pair is not admissible. Secondly, when the r/ξ ratio is below one, we calculate the number of expansion terms needed to have the error of calculation of the integral kernel less than a user prescribed criteria ϵ. If the number of expansion terms low enough, so that compression is achieved (equation (9)), this cluster pair is admissible. At this point the tree of pairs of clusters gets a leaf - no further branching is necessary.

Considering a cubic mesh of 16^3 cells with 33^3 nodes the matrix structure showing admissible and inadmissible blocks is shown on Figure 2.

The described FMM based algorithm was implemented into the BEM Poisson solver code. It is capable of constructing a data sparse approximation $[B']$ of the full domain matrix $[B]$ and use it to evaluate the right-hand side of the system of equations.

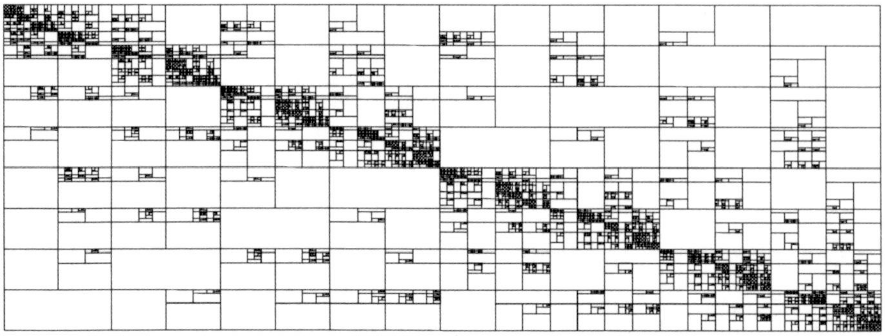

Figure 2: Matrix structure of a cubic mesh (16^3 cells, 33^3 nodes). Filled areas show inadmissible matrix blocks, white areas are admissible leafs obtained using an admissibility criteria of $\epsilon = 10^{-5}$. The corresponding tree of pairs of clusters had 19 levels.

4 Numerical tests

Let the problem domain be a unit cube. A structured mesh with equally spaced hexahedral cells fills the cube. Let N be the number of nodes in each direction. The number of nodes in the domain is N^3 and the number of nodes on the boundary scales as $\mathcal{O}(N^2)$. The number of elements in the fully populated right-hand side matrix $[B]$ thus scales as $\mathcal{O}(N^5)$. This fact is confirmed in Figure 3 where the storage requirements of the full matrix $[B]$ and the FMM compressed matrix

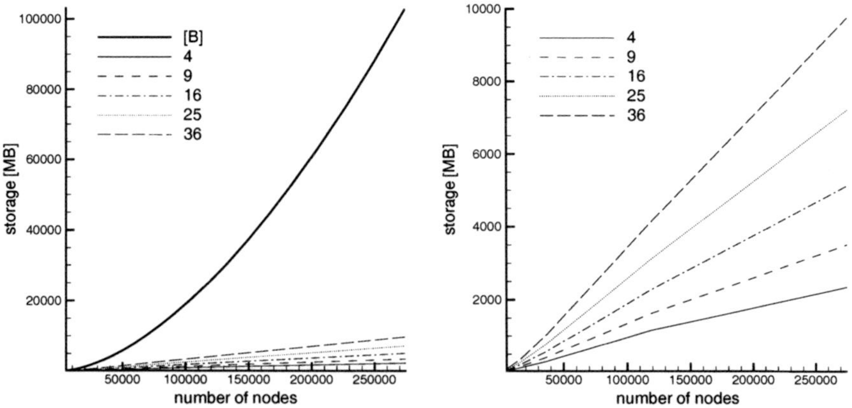

Figure 3: The graphs show a comparison of the memory required to store a full matrix $[B]$ (thick solid line) and FMM compressed matrices $[B']$ with different number of expansion terms. We observe the linear dependence of storage requirements for $[B']$ on the number of nodes regardless of the number of expansion terms.

$[B']$ are compared for different meshes. Naturally the storage requirements for $[B']$ increase when the number of expansion terms is increased. However looking at the storage requirement at a chosen number of expansion terms, we observe that it increases linearly with the number of nodes, i.e. scales as $\mathcal{O}(N^3)$. This relationship holds regardless of the number of expansion terms. Thus by employing the FMM we were able to decrease the storage requirements from $\mathcal{O}(n_d \cdot n_b)$ to linear dependence of $\mathcal{O}(n_d)$.

The accuracy of carrying out the calculation using the FMM domain matrix was analysed by solving the Poisson equations with known analytical solutions. The numerical solution was obtained with the use of the full matrix and by using the FMM data sparse domain matrix $[B']$. The problem geometry was a unit cube. Table 1 summarizes the equations and boundary conditions.

The uniform norms of solutions obtained using the full domain matrix $[B]$ are presented in Table 2. The same problems were solved using FMM $[B']$ matrices with different number of expansion terms. Let us define the data ratio $\mathcal{D}$ as the amount of data required to store the $[B']$ matrix divided by the amount of data in the $[B]$ matrix. A larger number of expansion terms results in larger $[B']$ matrix and thus a larger data ratio. Figures 4, 5 and 6 present uniforms norms for problems a), b) and c) respectively. We observe the norms increasing with decreasing data ratio. Comparing the results of the three problems, we see that the accuracy is best with problem a) and worst with problem c). This can be explained by the fact that the domain vector is a steeper function in problem c) than in problem a). At the same time we observe that the accuracy of solution converges with increasing data ratio to the accuracy full domain matrix solution (Table 2).

Table 1: Poisson equations with analytical solutions. The geometry was a unit cube. The boundary conditions in all cases were $u(x=0)=0$, $u(x=1)=1$, $q(y=0, y=1, z=0, z=1)=0$.

	Equation	Analytical solution
a)	$\nabla^2 u = 2$	$u_a = x^2$, $q_{a,x=0} = 0$, $q_{a,x=1} = 2$
b)	$\nabla^2 u = 6x$	$u_a = x^3$, $q_{a,x=0} = 0$, $q_{a,x=1} = 3$
c)	$\nabla^2 u = 12x^2$	$u_a = x^4$, $q_{a,x=0} = 0$, $q_{a,x=1} = 4$

Table 2: Solutions of Poisson equations in Table 1. Uniform norms for the full domain matrix $[B]$ solution against the analytical solution are presented.

	a)		b)		c)	
mesh	$\lVert u-u_a \rVert_\infty$	$\lVert q-q_a \rVert_\infty$	$\lVert u-u_a \rVert_\infty$	$\lVert q-q_a \rVert_\infty$	$\lVert u-u_a \rVert_\infty$	$\lVert q-q_a \rVert_\infty$
17^3	$6.2 \cdot 10^{-6}$	$6.4 \cdot 10^{-5}$	$2.4 \cdot 10^{-5}$	$7.7 \cdot 10^{-4}$	$9.0 \cdot 10^{-5}$	$2.9 \cdot 10^{-3}$
25^3	$2.8 \cdot 10^{-6}$	$4.0 \cdot 10^{-5}$	$7.4 \cdot 10^{-6}$	$3.4 \cdot 10^{-4}$	$2.8 \cdot 10^{-5}$	$1.3 \cdot 10^{-3}$
33^3	$1.6 \cdot 10^{-6}$	$3.9 \cdot 10^{-5}$	$3.1 \cdot 10^{-6}$	$1.9 \cdot 10^{-4}$	$1.2 \cdot 10^{-5}$	$6.8 \cdot 10^{-4}$

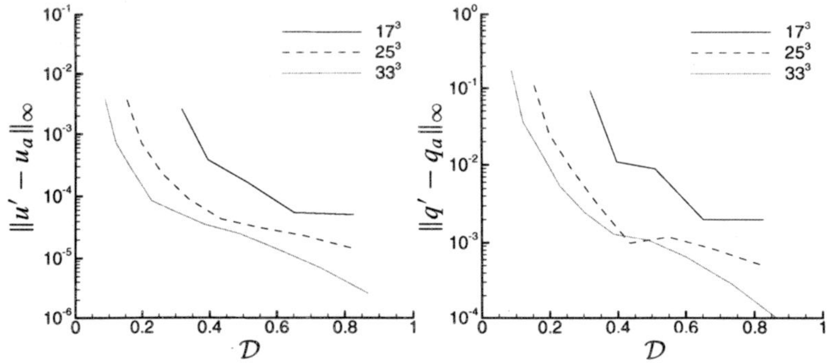

Figure 4: The graphs present uniform norms versus the data ratio $\mathcal{D}$ of solutions of problem a) in Table 1 using FMM matrices $[B']$.

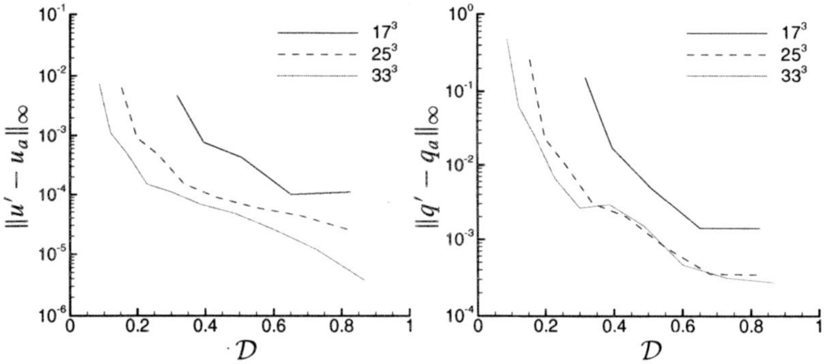

Figure 5: The graphs present uniform norms versus the data ratio $\mathcal{D}$ of solutions of problem b) in Table 1 using FMM matrices $[B']$.

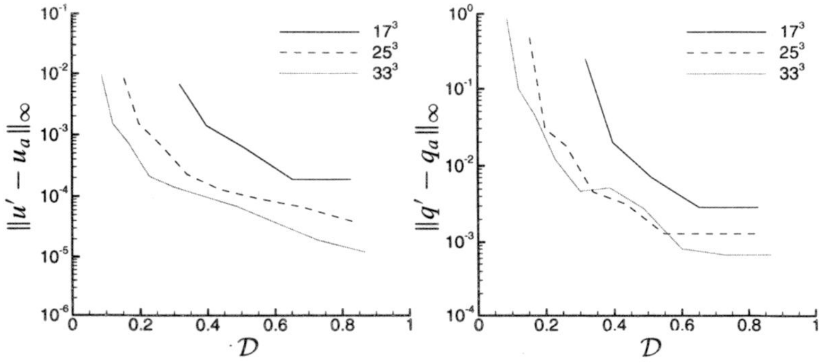

Figure 6: The graphs present uniform norms versus the data ratio $\mathcal{D}$ of solutions of problem c) in Table 1 using FMM matrices $[B']$.

5 Conclusions

We presented a Fast Multipole Method for data sparse representation and fast evaluation of the domain matrix of the BEM integral Poisson equation. The method uses a spherical harmonic series expansion of the integral kernel. The complexity of the calculation was decreased from $\mathcal{O}(n_d \cdot n_b)$ to linear dependence of $\mathcal{O}(n_d)$, where n_d is the number of nodes in the domain and n_b is the number of nodes on the boundary.

The method was tested on Poisson equations with known analytical solutions. It was shown, that for a given solution accuracy, the method enables better data ratios on denser meshes.

The main advantages of the method are the decreased storage requirements and the acceleration of matrix vector multiplication. Since only a small portion of the matrix must be evaluated, we observed a considerable acceleration of the time needed to set up the matrix as well.

In the near future, the method will be used to accelerate the solution of the kinematics equation, which is also of the Poisson type. The solution of this equation is an important part of the algorithm that solves fluid flow the velocity-vorticity formulation of incompressible Navier–Stokes equations.

References

[1] Ravnik, J., Škerget, L. & Žunič, Z., Combined single domain and subdomain BEM for 3D laminar viscous flow. *Eng Anal Bound Elem*, **submitted**, pp. xx–xx, 2008.

[2] Greengard, L. & Rokhlin, V., A fast algorithm for particle simulations. *J Comput Phys*, **73**, pp. 325–348, 1987.

[3] Bui, T.T., Ong, E.T., Khoo, B.C., Klaseboer, E. & Hung, K.C., A fast algorithm for modeling multiple bubbles dynamics. *J Comput Phys*, **216**, pp.430–453, 2006.

[4] Gumerov, N.A. & Duraiswami, R., Fast multipole method for the biharmonic equation in three dimensions. *J Comput Phys*, **215**, pp. 363–383, 2006.

[5] Popov, V. & Power, H., An O(N) Taylor series multipole boundary element method for three-dimensional elasticity problems. *Eng Anal Boud Elem*, **25**, pp. 7–18, 2001.

[6] Popov, V., Power, H. & Walker, S.P., Numerical comparison between two possible multipole alternatives for the BEM solution of 3D elasticity problems based upon Taylor series expansions. *Eng Anal Boun Elem*, **27**, pp. 521–531, 2003.

[7] Wrobel, L.C., *The Boundary Element Method*. John Willey & Sons Ltd, 2002.

[8] Barnes, J. & Hut, P., A hierarchical O(N log N) force calculation algorithm. *Nature*, **324**, pp. 446–449, 1986.

[9] Press, W.H., Teukolsky, S.A., Vetterling, W.T. & Flannery, B.P., *Numerical Recipes - The Art of Scientific computing, Second Edition*. Cambridge University Press, 1997.

Simulation of thin film flow around a cylinder on an inclined plane using the boundary element method

S. J. Baxter, H. Power, K. A. Cliffe & S. Hibberd
The University of Nottingham, UK

Abstract

Steady Stokes flow driven by gravity down an inclined plane wall around a circular cylinder attached to the wall is considered. The effects of the cylinder are examined for various flow configurations. Values for the unit normal and curvature of the free surface are found using a Hermitian radial basis function interpolation. All free surface profiles indicate an upstream peak, followed by a trough downstream of the obstacle with the peak decaying in a "horseshoe" shaped deformation. Flow profiles are governed by four parameters; the plane inclination, the Bond number, the contact angle and the obstacle geometry.
Keywords: BEM, three-dimensional, thin film, Newtonian viscous flow, cylinder.

1 Introduction

Thin film flows down an inclined plane wall driven by gravity can often be modelled as an incompressible Stokes flow. Flows usually involve interaction with an obstacle either fully submerged of protruding through the film surface.

Early published works considering two-dimensional film flows over obstacles utilized a variety of techniques for the numerical procedures required to obtain solutions. An overview of these publications is presented in Blyth and Pozrikidis [1].

Hayes et al. [2] considered a three-dimensional steady, thin, viscous liquid film down an inclined plane driven by gravity and over small topographies. The lubrication approximation was used as the basis for their model and formulates a single linear inhomogeneous evolution equation. The free surface shape was then obtained by formation of the appropriate Green's function.

WIT Transactions on Modelling and Simulation, Vol 47, © 2008 WIT Press
www.witpress.com, ISSN 1743-355X (on-line)
doi:10.2495/BE080021

The use of a direct boundary integral equation (BIE) for a three-dimensional Stokes flow analysis of film flow down an inclined plane and over an obstacle was first implemented by Pozrikidis and Thoroddsen [3] and later refined by Blyth and Pozrikidis [1]. The model in Pozrikidis and Thoroddsen [3] is based around the asymptotic limit where the size of a spherical obstacle is much smaller than the undisturbed film depth. Flows are shown to form an upstream peak and downstream trough with the peak decaying in the form of a "horseshoe". The extension of this work by Blyth and Pozrikidis [1] focuses on generalizing the earlier analysis to more arbitrarily sized obstructions. However, both simulations implement a linearization about the free surface deflection and thus the accuracy of significant deformations caused by large obstacles is unknown.

Numerical simulation of flow around obstacles has not been extensively considered. To the authors knowledge, the only applicable published work is Sellier [4]. This studies flow around a cylinder using the lubrication approximation, with the governing equations solved by the FEMLAB package. This approach is unable to impose a no slip boundary condition on the obstacle and is more relevant to far field displacements.

This paper uses the BIE formulation outlined by Blyth and Pozrikidis [1] to implement a modified analysis for film flows around large obstacles. A Hermitian radial basis function (RBF) is used for evaluation of the free surface terms required and is implemented as a replacement for previous finite difference approximations (FDAs) [1, 3]. Elimination of the restriction that the free surface deformation is small [1, 3], allows accurate results for large obstacles.

2 Mathematical formulation

Figure 1 shows a two-dimensional schematic of a typical film flow of undisturbed thickness H travelling down a plane inclined at angle α and around an obstacle attached to the plane. The inclined plane is referred to as S_w, the obstacle-fluid surface S_p the disturbed free surface S_f, and the outward unit normal is denoted

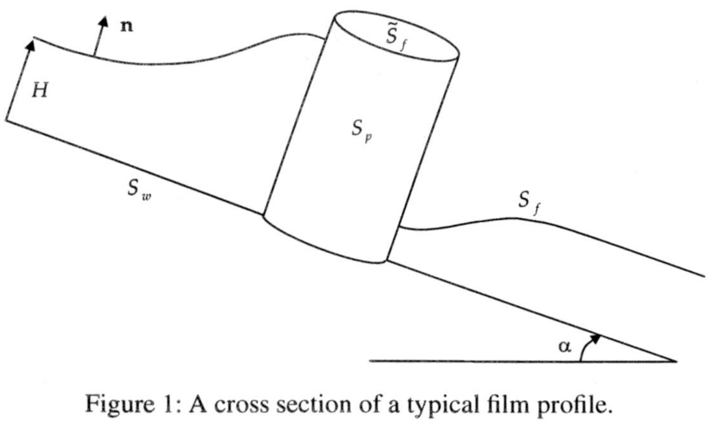

Figure 1: A cross section of a typical film profile.

n. Finally the Cartesian co-ordinate system is aligned so that the x_1x_2-plane defines the undisturbed surface, with x_1 in the direction of the flow, and x_3 perpendicular to the undisturbed flow.

For convenience the physical variables are first non-dimensionalized. The undisturbed film height, H is used for the reference length, the undisturbed free stream surface velocity, $U_s = \frac{H^2 \rho g \sin\alpha}{2\mu}$ (given by the Nusselt solution) for the reference velocity, and a viscous scaling $\frac{\mu U_s}{H}$ used for the tractions. The associated Bond number is given by,

$$Bo = \frac{\rho g H^2 \sin\alpha}{\gamma}, \tag{1}$$

where γ is the surface tension, ρ the fluid density and g the acceleration due to gravity. The inverse Bond number B is defined by Blyth and Pozrikidis [1] using the mass flux and is related to eqn (1) by $B = \frac{\sin^{1/3}\alpha}{Bo}$.

The flow problem is governed by the usual incompressible equations for conservation of mass eqn (2) and steady Stokes flow eqn (3),

$$\frac{\partial u_i}{\partial x_i} = 0, \tag{2}$$

$$-\frac{\partial p}{\partial x_i} + \frac{\partial G}{\partial x_i} + \frac{\partial^2 u_i}{\partial x_j^2} = 0, \tag{3}$$

where $G = -2(x_3 \cot\alpha - x_1)$. The far field boundary conditions involve the flow velocity and pressure returning to undisturbed values far from the obstacle and the free surface deflection decaying to zero, i.e,

$$\left.\begin{array}{r} u_i \to u_i^{\infty} \\ p \to p^{\infty} \\ h \to 0 \\ \dfrac{\partial h}{\partial x_1} \to 0 \\ \dfrac{\partial h}{\partial x_2} \to 0 \end{array}\right\} \text{as } \mathbf{x} \to \pm\infty. \tag{4}$$

Boundary conditions must include no slip (zero velocity) on the wall and obstacle surface eqn (5), and a kinematic eqn (6) and dynamic eqn (7) condition applied on the free surface. These requirements can be conveniently expressed as

$$u_i = 0 \quad \mathbf{x} \in S_w \cup S_p, \tag{5}$$

$$\frac{\partial x_i}{\partial t} n_i = u_j n_j \quad \mathbf{x} \in S_f, \tag{6}$$

$$f_i = \sigma_{ij} n_j = -\frac{4}{Bo} \kappa n_i \quad \mathbf{x} \in S_f, \tag{7}$$

where t is time, f_i is the boundary traction, κ is the curvature of the surface and σ_{ij} is the stress tensor. Expressions for curvature and the stress tensor are

$$\kappa = \frac{1}{2}\frac{\partial n_i}{\partial x_i}, \tag{8}$$

$$\sigma_{ij} = -p\delta_{ij} + \left(\frac{\partial u_i}{\partial x_j} + \frac{\partial u_j}{\partial x_i}\right). \tag{9}$$

Finally, a contact angle θ between the free surface and the obstacle boundary must be specified and dependent on the fluid / obstacle properties. This is incorporated into the problem by

$$\frac{\partial h}{\partial x_i} n_i = \tan\left(\frac{\pi}{2} - \theta\right), \tag{10}$$

which is valid along the contact line.

In the absence of an obstacle, the governing equations are

$$\frac{\partial u_i^\infty}{\partial x_i} = 0, \tag{11}$$

$$-\frac{\partial p^\infty}{\partial x_i} + \frac{\partial G}{\partial x_i} + \frac{\partial^2 u_i^\infty}{\partial x_j^2} = 0, \tag{12}$$

and u_i^∞, p^∞ corresponds to the far field asymptotic case. In addition the no-slip boundary conditions for the asymptotic flow give $u_i^\infty = 0$ for $\mathbf{x} \in S_w$, and solutions can be readily shown as

$$p^\infty = -2x_3 \cot\alpha, \tag{13a}$$

$$u_i^\infty = (1 - x_3^2)\delta_{i1}. \tag{13b}$$

The asymptotic boundary traction is given by $f_i^\infty = \sigma_{ij}(p^\infty, u_k^\infty)n_j$, and using eqn (9) becomes

$$f_i^\infty = 2x_3(\cot\alpha n_i - n_3\delta_{i1} - n_1\delta_{i3}). \tag{14}$$

Solutions eqn (13), eqn (14) govern the Stokes flow down an inclined plane in the absence of obstacles and are used as the basis for determining conditions for a disturbance flow due to the obstacle.

Disturbance velocities and pressures are denoted as $u_i^\delta = u_i - u_i^\infty$ and $p^\delta = p - p^\infty$ respectively. Governing equations, obtained by comparing eqns (2)–(3) with eqns (11)–(12) are

$$\frac{\partial u_i^\delta}{\partial x_i} = 0, \tag{15}$$

$$-\frac{\partial p^\delta}{\partial x_i} + \frac{\partial^2 u_i^\delta}{\partial x_j^2} = 0. \tag{16}$$

To satisfy the far-field condition eqn (4), u_i^δ and p^δ are subject to the conditions

$$\left.\begin{array}{l} u_i^\delta \to 0 \\ p^\delta \to 0 \end{array}\right\} \text{ as } \mathbf{x} \to \pm\infty. \tag{17}$$

Velocity conditions are found on the wall, S_w and obstacle, S_p and require

$$u_i^\delta = 0 \quad \mathbf{x} \in S_w, \tag{18a}$$

$$u_i^\delta = -u_i^\infty \quad \mathbf{x} \in S_p. \tag{18b}$$

In addition the associated disturbance boundary traction, f_i^δ is defined on the free surface S_f from eqn (7) as

$$f_i^\delta = -\frac{4}{Bo}\kappa n_i - f_i^\infty, \tag{19}$$

where f_i^∞ is given by eqn (14), and κ is the curvature associated with the disturbed free surface.

2.1 Boundary integral formulation

The equations of Stokes flow for the disturbance quantities eqns (15)–(16) can be expressed exactly as an integral representation eqn (20),

$$\begin{aligned} c_{ij}(\mathbf{x}_0)u_i^\delta(\mathbf{x}_0) = & \frac{1}{8\pi}\int_{S_f \cup S_p} G_{ij}^*(\mathbf{x}, \mathbf{x}_0) f_i^\delta(\mathbf{x}) \mathrm{d}S(\mathbf{x}) \\ & - \frac{1}{8\pi}\int_{S_f \cup S_p} u_i^\delta(\mathbf{x}) T_{ijk}^*(\mathbf{x}, \mathbf{x}_0) n_k(\mathbf{x}) \mathrm{d}S(\mathbf{x}). \end{aligned} \tag{20}$$

To close the obstacle a virtual surface is introduced and labeled $\tilde{S}_f$. For the closed domain $S_p \cup \tilde{S}_f$, the Stokes flow associated with the far field quantities u_i^∞ and $p^\infty - G$ is satisfied by the integral representation eqn (21)

$$\begin{aligned} c_{ij}(\mathbf{x}_0)u_i^\infty(\mathbf{x}_0) = & -\frac{1}{8\pi}\int_{S_p \cup \tilde{S}_f} G_{ij}^*(\mathbf{x}, \mathbf{x}_0)(f_i^\infty(\mathbf{x}) + G n_i(\mathbf{x})) \mathrm{d}S(\mathbf{x}) \\ & + \frac{1}{8\pi}\int_{S_p \cup \tilde{S}_f} u_i^\infty(\mathbf{x}) T_{ijk}^*(\mathbf{x}, \mathbf{x}_0) n_k(\mathbf{x}) \mathrm{d}S(\mathbf{x}). \end{aligned} \tag{21}$$

The coefficient $c_{ij}(\mathbf{x}_0)$ takes a value 0 when $\mathbf{x}_0$ is outside the domain, $\frac{1}{2}\delta_{ij}$ when $\mathbf{x}_0$ is on the domain boundary and δ_{ij} when $\mathbf{x}_0$ is within the domain. Functions $G_{ij}^*(\mathbf{x}, \mathbf{x}_0)$, $T_{ijk}^*(\mathbf{x}, \mathbf{x}_0)$ are the Lorentz-Blake Green's function for velocity and stress respectively satisfying zero velocity value at the inclined plane. Noting that $f_i^\delta \to 0$ and $u_i^\delta \to 0$ as $\mathbf{x} \to \pm\infty$, the edges of the flow domain are omitted from eqn (20) as well.

By a similar derivation to flow over an obstacle as described by Blyth and Pozrikidis [1], two BIE are derived, one collocated over the free surface S_f and one over the obstacle-fluid surface S_p. The BIE for $\mathbf{x}_0 \in S_f$ is,

$$\begin{aligned}
&\frac{1}{2}u_j^\delta(\mathbf{x}_0) + \frac{1}{8\pi}\int_{S_f} u_i^\delta(\mathbf{x})T_{ijk}^*(\mathbf{x}, \mathbf{x}_0)n_k(\mathbf{x})\mathrm{d}S(\mathbf{x}) \\
&\quad = \frac{1}{8\pi}\int_{S_p} G_{ij}^*(\mathbf{x}, \mathbf{x}_0)\tilde{f}_i(\mathbf{x})\mathrm{d}S(\mathbf{x}) \\
&\quad + \frac{1}{8\pi}\int_{S_f} G_{ij}^*(\mathbf{x}, \mathbf{x}_0)f_i^\delta(\mathbf{x})\mathrm{d}S(\mathbf{x}) \\
&\quad + \frac{1}{8\pi}\int_{\tilde{S}_f} G_{ij}^*(\mathbf{x}, \mathbf{x}_0)(f_i^\infty(\mathbf{x}) + Gn_i(\mathbf{x}))\mathrm{d}S(\mathbf{x}) \\
&\quad - \frac{1}{8\pi}\int_{\tilde{S}_f} u_i^\infty(\mathbf{x})T_{ijk}^*(\mathbf{x}, \mathbf{x}_0)n_k(\mathbf{x})\mathrm{d}S(\mathbf{x}) \quad \mathbf{x}_0 \in S_f,
\end{aligned} \tag{22}$$

where $\tilde{f}_i(\mathbf{x}) = f_i(\mathbf{x}) + Gn_i$. For $\mathbf{x}_0 \in S_p$ the BIE is derived as

$$\begin{aligned}
\frac{1}{8\pi}\int_{S_p} G_{ij}^*(\mathbf{x}, \mathbf{x}_0)\tilde{f}_i(\mathbf{x})\mathrm{d}S(\mathbf{x}) &= -u_j^\infty(\mathbf{x}_0) - \frac{1}{8\pi}\int_{S_f} G_{ij}^*(\mathbf{x}, \mathbf{x}_0)f_i^\delta(\mathbf{x})\mathrm{d}S(\mathbf{x}) \\
&+ \frac{1}{8\pi}\int_{S_f} u_i^\delta(\mathbf{x})T_{ijk}^*(\mathbf{x}, \mathbf{x}_0)n_k(\mathbf{x})\mathrm{d}S(\mathbf{x}) \\
&- \frac{1}{8\pi}\int_{\tilde{S}_f} G_{ij}^*(\mathbf{x}, \mathbf{x}_0)(f_i^\infty(\mathbf{x}) + Gn_i(\mathbf{x}))\mathrm{d}S(\mathbf{x}) \\
&+ \frac{1}{8\pi}\int_{\tilde{S}_f} u_i^\infty(\mathbf{x})T_{ijk}^*(\mathbf{x}, \mathbf{x}_0)n_k(\mathbf{x})\mathrm{d}S(\mathbf{x}) \quad \mathbf{x}_0 \in S_p.
\end{aligned} \tag{23}$$

In summary the steady, gravity driven Stokes flow down an inclined plane around an obstacle is governed exactly by solutions satisfying the BIE's given in eqn (22) and eqn (23), the kinematic condition eqn (6) and the dynamic condition eqn (7), along with the far field equations eqn (4). As the obstacle protrudes through the free surface a contact line is present and the contact angle condition eqn (10) is also required.

2.2 Finite free surface deflections and the radial basis functions

A modification to the linearized approach of Blyth and Pozrikidis [1] using a Hermitian RBF allows accurate calculations of the free surface curvature and spatial derivatives. Further, the method also allows the incorporation of the free surface far field conditions and the contact angle condition. At any point on the free surface, the height is given by $h(x_1, x_2)$, and takes values h^i at data point i for a total of N points. The surface can then be interpolated using a RBF $\psi(\|\mathbf{x}-\boldsymbol{\xi}^j\|)$ where $j = 1, \ldots, N$ and $\boldsymbol{\xi}^j$ are the x_1, x_2 co-ordinates of the data points h^j. Thin plate splines (TPS) was used for the radial basis function, and require the addition of a polynomial term of order $m - 1$ to guarantee the invertibility of the interpolation matrix, where m is specified within the thin plate spline.

The values of $\frac{\partial h}{\partial x_1}$ and $\frac{\partial h}{\partial x_2}$ can be constrained at a selected set of n data points, whilst still defining h on the original N data points. A solution for h can be obtained using the differential operators $\frac{\partial}{\partial x_1}$ and $\frac{\partial}{\partial x_2}$ but acting on the thin plate splines second argument $\boldsymbol{\xi}$, i.e. $\frac{\partial}{\partial \xi_1}$ and $\frac{\partial}{\partial \xi_2}$. Similar constraint conditions are imposed in the interpolation to satisfy the contact condition eqn (10).

To complete the computation for finite disturbances, the outward unit normal and mean curvature of the free surface are both required. Evaluation of the unit normal will require derivatives $\frac{\partial h}{\partial x_1}$ and $\frac{\partial h}{\partial x_2}$, with the mean curvature needing in addition, $\frac{\partial^2 h}{\partial x_1^2}$, $\frac{\partial^2 h}{\partial x_2^2}$, and $\frac{\partial^2 h}{\partial x_1 \partial x_2}$ on the surface, which are obtained by the radial basis interpolation.

3 Numerical examples

Figure 2 shows the obtained numerical result of the free surface mesh for flow around a cylinder of radius $a = 2.0$. The flow was down a plane inclined at $\alpha =$

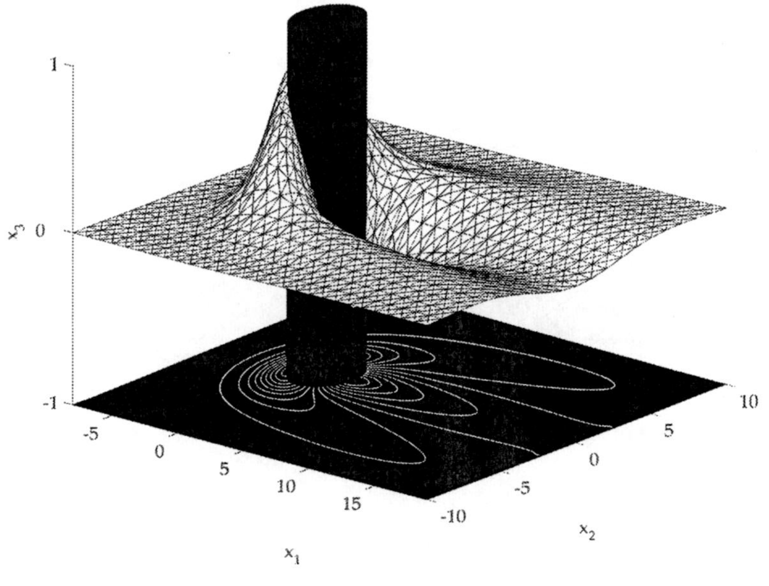

Figure 2: The free surface mesh and contours showing flow around a large cylinder of radius $a = 2.0$. The plane is inclined at $\alpha = 45°$, the contact angle is set to $\theta = 90°$ and the inverse Bond number is 1.

$45°$ and the film has an inverse Bond number of $B = 1$. The contact angle at the cylinder walls was constrained to $\theta = 90°$.

Variations of the inverse Bond number B was considered for flow down a plane inclined at $\alpha = 45°$ around a cylindrical obstacle of radius $a = 0.5$. The free surface - obstacle interface was modelled by a contact angle of $\theta = 90°$. Increasing the inverse Bond number, corresponding to a raising of surface tension forces, was shown to flatten the centre line profiles. Variations of plane inclination angles α were also considered. Flow with an inverse Bond number $B = 1$ and around a cylindrical obstacle of radius $a = 0.5$ was considered. The free surface - obstacle interface was modelled by a contact angle of $\theta = 90°$. Increasing the inclination angle resulted in a raising of the peak before the obstacle.

Figure 3 indicates the smoothed centre line solutions for flow around a range of cylinder radii. Flow is for an inverse Bond number $B = 1$ and down a plane inclined at $\alpha = 45°$. The obstacle was chosen neutrally wetting, (i.e. the contact angle was $\theta = 90°$). Increasing the cylinder radius results in a raising of the peak and deepening of the trough heights both before and behind the obstacle. Figure 4 illustrates the contact line profile around the cylinder wall for the range of cylinder radii depicted in figure 3. Interestingly intersection of the contact line profiles occurs at an approximately fixed angular position along the cylinder, between 0.94–0.98 radians upstream of the position $x_1 = a$.

Figure 5 indicates the smoothed centre line solutions for flow around a cylinder of radius 0.5. Flow is for an inverse Bond number $B = 1$ and down a plane

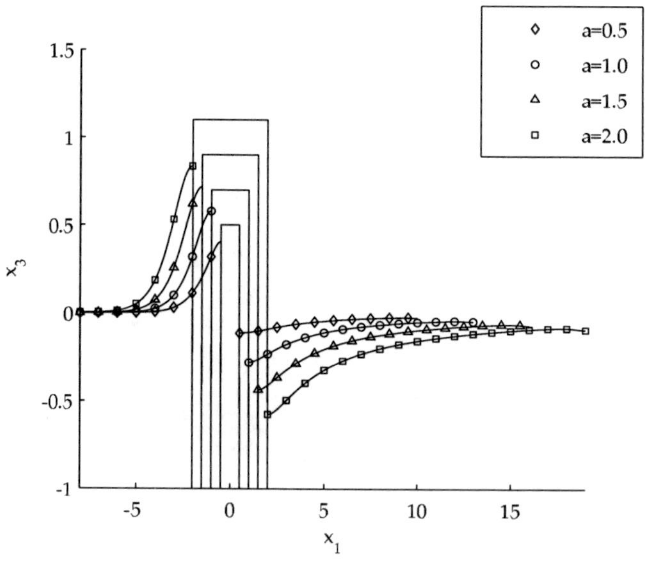

Figure 3: Centre line solutions for free surface deflections around a cylinder of varying radius, with inverse Bond number $B = 1$ and down a plane inclined at $45°$. The contact angle at the cylinder wall is $\theta = 90°$.

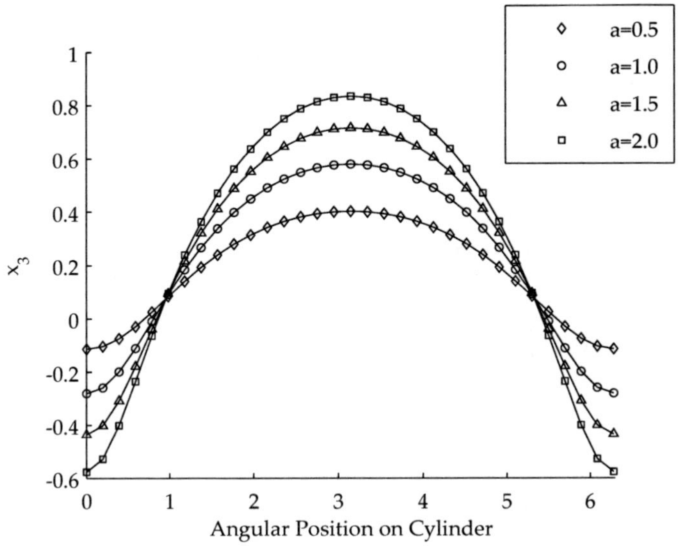

Figure 4: Contact line solutions for free surface deflections around a cylinder of varying radius, with inverse Bond number $B = 1$ and down a plane inclined at 45°. The contact angle at the cylinder wall is $\theta = 90°$.

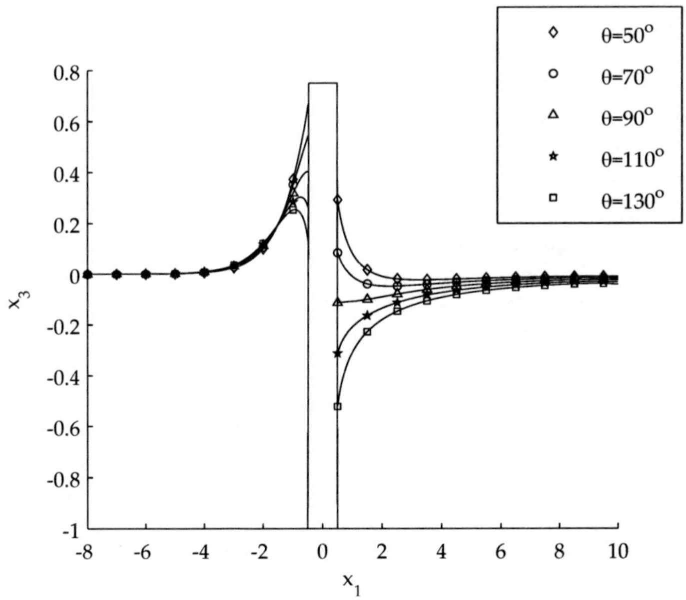

Figure 5: Centre line solutions for various contact angles at the cylinder wall. The free surface deflections are around a cylinder of radius $a = 0.5$ on an inclined plane at 45° and the flow has an inverse Bond number $B = 1$.

inclined at $\alpha = 45°$. The solutions show a range of wetting and non-wetting flow configurations corresponding to a range of contact angles θ. Decreasing the contact angle results in a raising of the peak and trough heights both before and behind the obstacle.

4 Conclusions

Stokes flow down an inclined plane around a cylinder is considered and solutions found by the Boundary Element Method (BEM). Use of a global Radial Basis Function (RBF) is used to extend the small free surface deflection assumption by Blyth and Pozrikidis [1] through accurate determination of free surface quantities such as curvature and unit normal for larger surface deflections. For flow around cylinders incorporation of a contact angle condition within the RBF was required, and within a parameter study results for various size cylinders presented.

Acknowledgements

This research was carried out at the University Technology Centre in Gas Turbine Transmission Systems at the University of Nottingham with financial support from Rolls-Royce plc, Aerospace Group. The views expressed in this paper are those of the authors and not necessarily those of Rolls-Royce plc, Aerospace Group.

References

[1] Blyth, M.G. & Pozrikidis, C., Film flow down an inclined plane over a three-dimensional obstacle. *Physics of Fluids*, **18(5)**, pp. 052104–14, 2006.

[2] Hayes, M., O'Brien, G. & Lammers, J.H., Greens function for steady flow over a small two-dimensional topography. *Physics of Fluids*, **12(11)**, pp. 2845–2858, 2000.

[3] Pozrikidis, C. & Thoroddsen, S.T., The deformation of a liquid film flowing down an inclined plane wall over a small particle arrested on the wall. *Physics of Fluids A*, **3(11)**, pp. 2546–2558, 1991.

[4] Sellier, M., The flow of a thin liquid film past a cylinder. *COMSOL Users Conference 2006*, Frankfurt, pp. 151–154, 2006.

Meshless and BE analysis of the flow and concentration around a DO-supplying machine

K. Sakamoto[1], M. Kanoh[2] & T. Kuroki[3]
[1]*Environment Division, Matsue Doken Co. Ltd., Japan*
[2]*Department of Civil Engineering, Kyushu Sangyo University, Japan*
[3]*Baikoen 1-15-21 Dazaifu, Fukuoka 818-0124, Japan*

Abstract

In an earlier study, a meshless method was developed to obtain stable and convergent results for the flow of the entire area of a water reservoir. Shortage or lack of dissolved oxygen (DO) sometimes results in pollution of the water in a reservoir. An attempt was made to ameliorate the concentration of oxygen in the lower layer of the reservoir by using a machine that supplies DO. A meshless method was applied to calculate the concentration distribution of DO and the flow caused by the DO-supplying machine. For the area surrounding the DO-supplying machine, the results calculated by the meshless method appeared to be slightly unreasonable compared with the observed velocity vectors obtained in the model simulation of a water reservoir constructed in our laboratory. The boundary element method did not succeed in obtaining convergent solutions to this problem. In this study, a new meshless method is devised, and our boundary element method is improved so that analytical solutions that are comparable with the observed results can be obtained. Regarding the concentration distribution of DO and the velocity vectors of the flow calculated by the two numerical methods (the WFDM and the FEM) and those observed in our model simulation described above, the effect and accuracy of the alternative meshless and boundary element methods were estimated.

Keywords: meshless method, boundary element method (BEM), flow and concentration in water reservoirs, weighted finite difference method (WFDM), finite element method (FEM) and observed velocity in model simulation of water reservoir.

doi:10.2495/BE080031

1 Introduction

By using a machine that supplies DO (dissolved oxygen), we attempted to ameliorate the concentration of oxygen in the lower layer of water areas, such as a water reservoir or the sea (Kanoh et al. [1]), and succeeded in several cases. Our DO-supplying machine dissolved the high concentration of oxygen (about 100 mg per litre: [mg/L]) in the water of the tank of the machine at a depth of about 50 metres. The water that became rich in DO flowed out to the lower layer of the water area in a horizontal direction. The observed velocity of the water flow was considerably small (almost 0.1 m/sec [metre per second]); however, the distance reached by the DO-rich water was observed to be more than 300 metres in the B reservoir. No theoretical explanation or evidence has been found to explain these phenomena. In order to obtain the evidence to explain the phenomena, we constructed a new simulation model in the sanitary and environmental engineering laboratory of Kyushu Sangyo University and obtained some observed velocity vectors and the distributions of the DO concentration in the model. Our meshless and boundary element methods were applied to numerically simulate the phenomena and compared with the observed data obtained in our model simulation.

2 Governing equations

Three equations, i.e., continuous, Navier-Stokes (N-S), and convective-diffusion equations, govern the flow and diffusion in a water reservoir. In the vertical (x_1, x_2) plane, as illustrated in Figure 1, these equations are shown as follows:

$$u_{1,1} + u_{2,2} = 0 \tag{1}$$

$$\dot{u}_1 + u_1 \cdot u_{1,1} + u_2 \cdot u_{1,2} - \nu(u_{1,,1} + u_{1,,2}) = -P_{,1}/\rho \tag{2$_1$}$$

$$\dot{u}_2 + u_1 \cdot u_{2,1} + u_2 \cdot u_{2,2} - \nu(u_{2,,1} + u_{2,,2}) = -P_{,2}/\rho + g \tag{2$_2$}$$

$$\dot{C} + u_1 \cdot C_{,1} + u_2 \cdot C_{,2} - D_1 \cdot C_{,,1} - D_2 \cdot C_{,,2} = 0 \tag{3}$$

$$\rho = A \cdot C + B \tag{4}$$

where u_1 and u_2 describe the velocities of the x_1 and x_2 directions, respectively, P is the pressure, g is the gravity acceleration, ν is the kinematic viscosity, C is the concentration of dissolved oxygen (DO), and D_1 and D_2 are the diffusion coefficients of the x_1 and x_2 directions, respectively. Here, the density ρ is connected to the DO concentration C, as written in Equation (4), with the coefficients A and B. The DO concentration C is compatible with the water temperature T in case it is necessary to calculate the water temperature T or other values.

3 Application of the meshless method

3.1 Meshless method for flow analysis

We tried to apply the meshless method to a flow analysis in the steady state in the area surrounding a DO-supplying machine in a water reservoir, as shown in

Figures 1(a) and (b). The meshless method is based on the idea of the mesh-free RBF collocation method (Kanoh et al. [1]). The penalty method was used so that the pressure terms would be eliminated in the N-S equations and the difficulty of the pressure boundary conditions would be avoided in the meshless method.

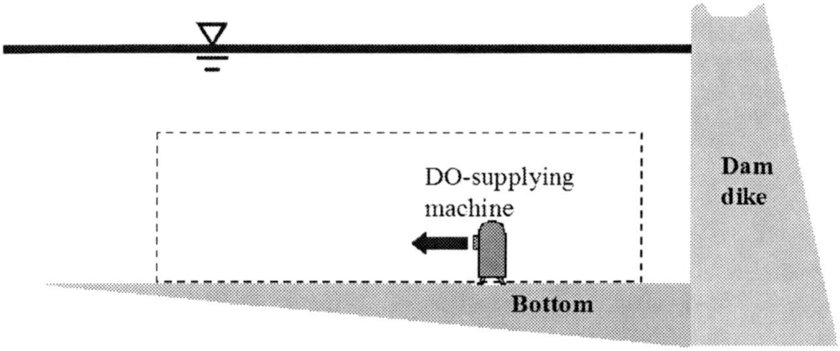

(a)

(b)

Figure 1: (a) Concept of a DO-supplying machine in a water reservoir. (b) Analytical domain around a DO-supplying machine.

3.2 Meshless method formulation for flow analysis

3.2.1 Application of the penalty method to N-S equations

Introducing the penalty method to the N-S equations in the steady state, the following equation is obtained, as described below,

$$u_j \cdot u_{i,j} - \lambda(u_{1,2} + u_{2,2})_{,j} \cdot \delta_{i,j} - \nu(u_{1,2} + u_{2,2})_{,j} = 0 \quad i, j = 1, 2 \tag{5}$$

where λ means $R_e \cdot K/\rho$, R_e is the Reynolds number, and K describes the coefficient of the penalty method (Kanoh et al. [1, 2]).

3.2.2 Simultaneous equations for the meshless method of flow analysis

Substituting $\underline{u}_1$ and $\underline{u}_2$ at time (t-Δt) into the above Equation (5), the following expression is obtained:

$$\underline{u}_1 \cdot u_{1,1} + \underline{u}_2 \cdot u_{1,2} - \lambda(u_{1,11} + u_{2,21}) - \nu(u_{1,22} + u_{2,12} + 2u_{1,11}) = 0 \tag{6$_1$}$$

$$\underline{u}_1 \cdot u_{2,1} + \underline{u}_2 \cdot u_{2,2} - \lambda(u_{1,12} + u_{2,22}) - \nu(u_{2,11} + u_{1,21} + 2u_{2,22}) = 0 \tag{6$_2$}$$

The unknown values u_1 and u_2 are expressed as Equation (7) using the global expansion function Xj (= $(r^2+c^2)^{-1/2}$) of the mesh-free RBF collocation method,

$$u_1 = \alpha_j Xj = \alpha_j/(r^2+c^2)^{1/2} \quad (7_1)$$

$$u_2 = \beta_j Xj = \beta_j/(r^2+c^2)^{1/2} \quad (7_2)$$

where r equals $\{(x-x_j)^2+(y-y_j)^2\}^{1/2}$ and c is the constant.

Solving the simultaneous equations that are described as Equation (8), the above unknowns (α_j and β_j) can be obtained.

$$\left\{\left(\underline{u}_1\frac{\partial X_j}{\partial x_1}+\underline{u}_2\frac{\partial X_j}{\partial x_2}\right)-(\nu+\lambda)\frac{\partial^2 X_j}{\partial x_1{}^2}-\nu(\frac{\partial^2 X_j}{\partial x_1{}^2}+\frac{\partial^2 X_j}{\partial x_2{}^2})\right\}\alpha_j-(\nu+\lambda)\frac{\partial^2 X_j}{\partial x_1 \partial x_2}\beta_j=0 \quad (8_1),$$

$$-(\nu+\lambda)\frac{\partial^2 X_j}{\partial x_1 \partial x_2}\alpha_j+\left\{\left(\underline{u}_1\frac{\partial X_j}{\partial x_1}+\underline{u}_2\frac{\partial X_j}{\partial x_2}\right)-(\nu+\lambda)\frac{\partial^2 X_j}{\partial x_2{}^2}-\nu(\frac{\partial^2 X_j}{\partial x_1{}^2}+\frac{\partial^2 X_j}{\partial x_2{}^2})\right\}\beta_j=0 \quad (8_2).$$

Substituting the obtained values of α_j and β_j into Equation (7), the values of u_1 and u_2 in the steady state can be calculated using the meshless method.

3.3 Meshless method formulation for concentration analysis

Here, we deal with the concentration analysis in the unsteady state in the area surrounding the DO-supplying machine. First of all, the concentration in the steady state is expressed as Equation (9) with Equation (10) (Kanoh et al. [1]).

$$C = \gamma_j Xj = \beta_j/(r^2+c^2)^{1/2} \quad (9)$$

$$\left\{\left(\underline{u}_1\frac{\partial X_j}{\partial x_1}+\underline{u}_2\frac{\partial X_j}{\partial x_2}\right)-(D_1\frac{\partial^2 X_j}{\partial x_1{}^2}+D_2\frac{\partial^2 X_j}{\partial x_2{}^2})\right\}\gamma_j=0 \quad (10).$$

Secondly, the transient convective-diffusion equation is rewritten as follows:

$$\dot{C} + L(C) = 0 \quad (11),$$

where $\dot{C}$ is the time derivative of C and L(C) has the terms of convection and diffusion in the steady state. Thirdly, applying the finite difference scheme, Equation (11) yields

$$(C^{t+\Delta t} - C^t)/\Delta t + \{L^{t+\Delta t}(C) + L^t(C)\}/2 = 0 \quad (12),$$

$$C^{t+\Delta t} + L^{t+\Delta t}(C)\cdot\Delta t/2 = C^t - L^t(C)\cdot\Delta t/2 \quad (13),$$

where $C^{t+\Delta t}$ and C^t are the concentrations at time (t+Δt) and time (t), respectively, and $L^{t+\Delta t}$ (C) and L^t (C) are the terms of convection and diffusion at time (t+Δt) and time (t), respectively. Finally, using equations (9), (10), and (13), the meshless method can analyse the DO concentration in the unsteady state.

3.4 Boundary conditions and boundary discretisation

The boundary conditions for the flow and concentration analyses have been previously proposed for the meshless method, the BEM, the FEM, and the WFDM (Kanoh et al. [1–3]). Regarding the boundary discretisation for the flow analysis by our boundary element method, the outline is as follows: (1) the convergence and accuracy of the very delicate flow were sensitive to the domain and boundary discretisation. It is necessary to determine the appropriate lengths of the domain and boundary discretisation to obtain convergence and accuracy in the computational analysis; (2) in an earlier study, in which several kinds of lengths of the boundary elements were used for flow analysis around the DO-supplying machine, the boundary element method could not obtain the convergence in the analysis; (3) in this paper, two sub-regions and same-length elements are used, and inner cells are also introduced to calculate the volume integration. The new boundary element method, in which two kinds of lengths of boundary elements are employed (160 elements, 800 inner cells; and 800 elements, 1,890 inner cells), can obtain both the convergence and the comparable computational time with the meshless method or the FEM.

4 Results and discussion

As described above, we constructed a new simulation model in our laboratory and obtained some observed velocity vectors and the distributions of the DO concentration in the model. In reference to the observed results, we tried to obtain some evidence to explain the phenomena that the distance reached by the DO-rich water was more than 300 metres in a reservoir in spite of the small velocity of the water flowing out. The numerical results of the meshless method, the boundary element method (BEM), the finite element method (FEM), and the weighted finite difference method (WFDM) are also discussed in this section in order to obtain some evidence to explain the phenomena. The two kinds of analyses conducted using these methods are described here. Namely, the first analysis is that of the flow, and the second is that of the DO concentration around a DO-supplying machine in a water reservoir.

4.1 Observed values in a model around a DO-supplying machine

4.1.1 Flow velocity in a model

Figure 2 is an illustration of the velocity vectors caused by a DO-supplying machine in a reservoir model visualised using aluminium flakes, a strobe light, and a digital VTR. In the area neighbouring the outflow point, the velocity vectors in the horizontal direction were distinguished, and the values of the velocities were larger than those in other areas. In other areas, which were toward the centre or more than 0.8 metres from the outflow point, the velocity values became less than 1cm/sec.

4.1.2 Concentration distribution of DO in a model

Figure 3 is an illustration of the concentration distribution caused by a DO-supplying machine in a reservoir model visualised using a pigment (methylene blue) and a VTR. The speed of diffusion of the DO-rich water (DO: 30mg/L) was obviously quicker than that in DO-poor water (DO: 10mg/L; the figure was omitted in this paper), and the observed direction of the diffusion was mainly horizontal.

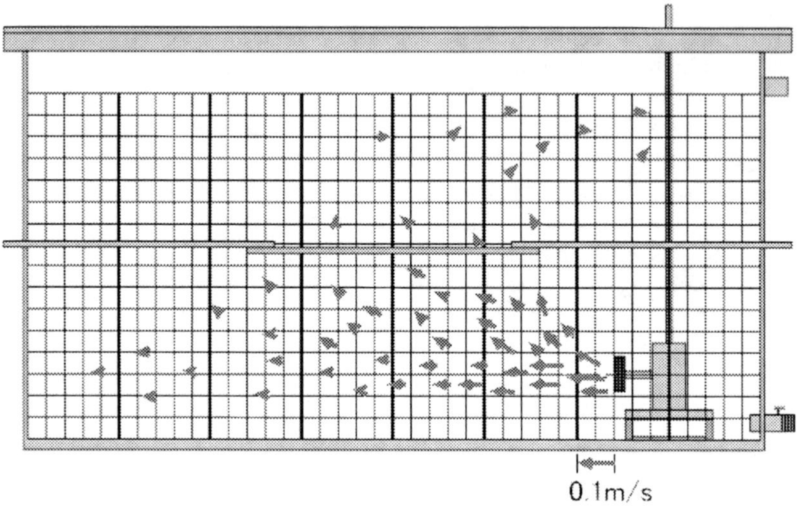

Figure 2: Observed velocity vector distribution (DO: 30mg/L).

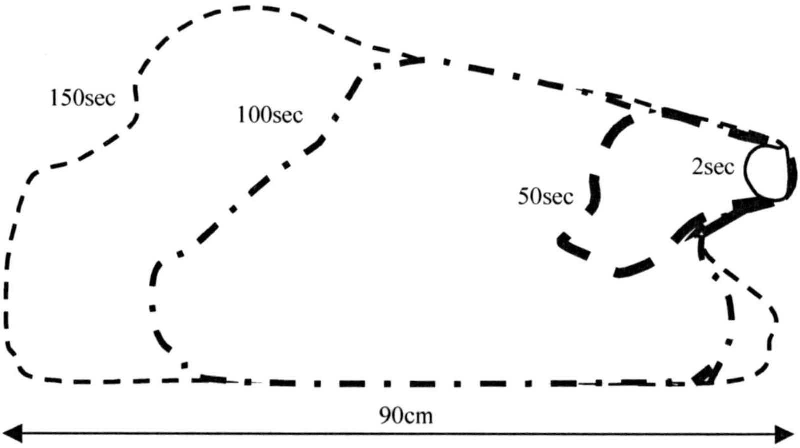

Figure 3: Observed areas of diffusion of pigment (DO: 30mg/L).

4.2 Flow analysis around a DO-supplying machine in a water reservoir

4.2.1 Meshless method calculation of the flow around the machine

Figure 4 is an illustration of the velocity vectors around a DO-supplying machine calculated by the meshless method, in which the number of the points in the meshless method is 1,891 and the three values of λ, C, and ν are 1,000.0, 1.0, and 0.001, respectively. The stability and convergence of the flow analysis around the machine to supply DO using the meshless method seemed satisfactory.

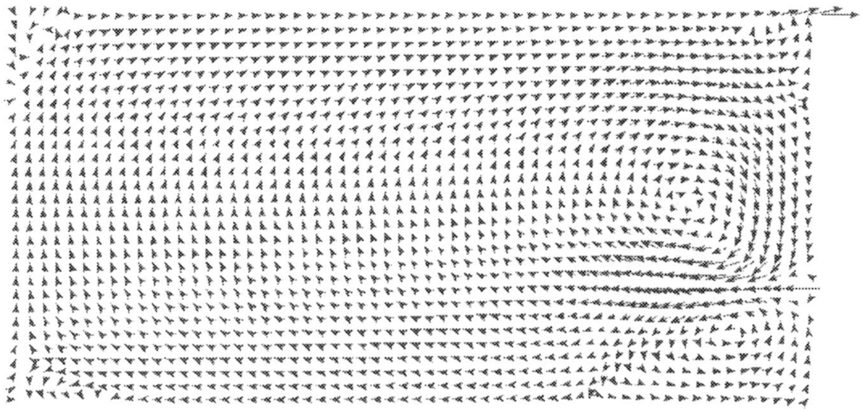

Figure 4: Velocity vectors using meshless method (Number of the points:1891).

4.2.2 BEM, FEM, and WFDM calculation of the flow around the machine

Figure 5 is an illustration of the velocity vectors around a DO-supplying machine calculated by the BEM, in which the number of elements in the BEM is 240 and the two values of λ and ν are 1,000.0 and 0.001, respectively. The penalty

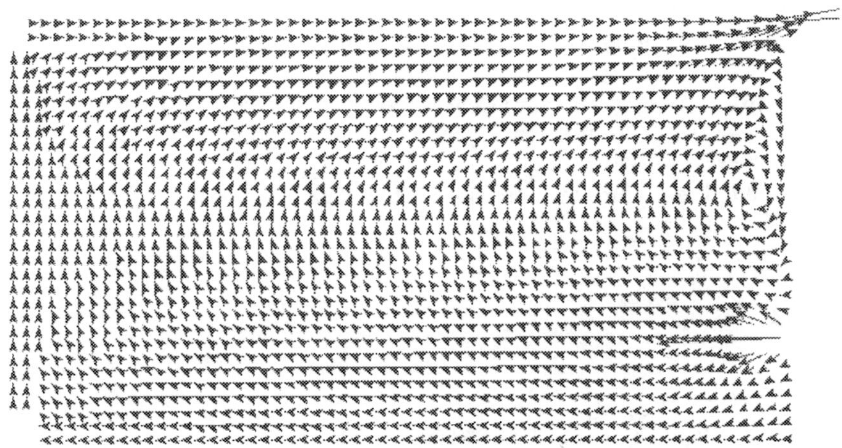

Figure 5: Velocity vectors using the BEM around a DO-supplying machine.

method was introduced, as well as in the meshless method. The stability and convergence of the flow analysis using the BEM seemed satisfactory. The stability and convergence of the flow analysis using both the FEM and the WFDM also seemed satisfactory (the figures to illustrate those velocity vectors have been omitted). These three methods could yield similar solutions to the true results of several flow problems that were observed in simulation models constructed in our laboratory.

4.2.3 Time required by the three numerical methods for the flow analysis

Table 1 shows the time required by the three numerical methods for analysing the flow around a DO-supplying machine. When the number of the divisions of the analytical domain was 800, the BEM and the meshless method needed almost 7 and 21 times the time required for the FEM, respectively. When the number of the divisions of the analytical domain became 1,800, the BEM and the meshless method needed almost 6 and 82 times the time required for the FEM, respectively. In both cases, for the purposes of saving time, the FEM was the best; the BEM was second best; and the meshless method was the poorest performer. On the other hand, for the purpose of saving the time and labour required for preparing the input data, the meshless method was the best, FEM was second best, and BEM was the worst.

Table 1: The time required by the three methods for analysing the flow around a DO-supplying machine.

Relative computational time / Numerical method	Number of divisions	
	800	1,600
FEM	1.00	5.00
BEM	6.67	28.3
Meshless method	20.7	413

4.3 Concentration analysis around a DO-supplying machine

4.3.1 Meshless method calculation of the concentration distribution

Figure 6 is an illustration of the concentration distribution calculated using the meshless method, in which the number of the points in the meshless method is 1,891 and the three values of λ, C, and ν are 1000.0, 2.2, and 0.001, respectively.

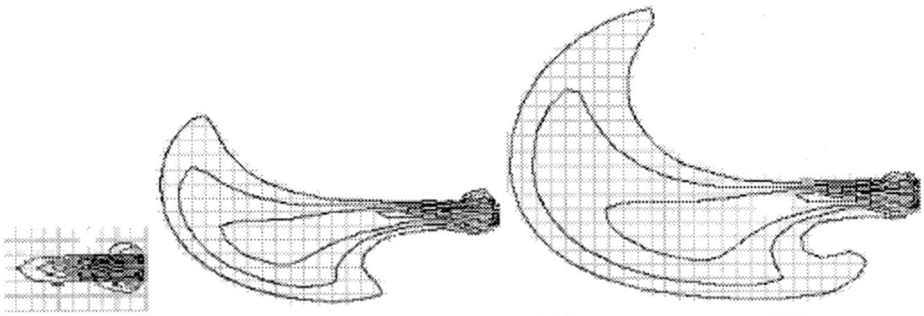

t = 4 sec t = 100sec t = 200sec

Figure 6: DO-concentration distribution using the meshless method (the contours were drawn when C was larger than 12mg/L).

We consider that the stability and convergence of the meshless method for this problem are satisfactory.

4.3.2 WFDM calculation of the concentration distribution

Figure 7 is an illustration of the concentration distribution calculated using the weighted finite difference method, in which the number of meshes in the WFDM is 12,960 and the value of ν is 0.001. We consider that the convergence and accuracy of the WFDM for this problem are satisfactory enough. In Figure 3, the observed concentration distribution (DO: 30mg/L) was illustrated in the former section. Comparing the observed values with the calculated results using the WFDM, it was evident that the WFDM could yield a similar solution to the actual results of the problem. The time required by the WFDM for analysing the concentration was less than those by the FEM, the BEM, and the meshless method, since the scheme of the WFDM was explicit and did not require any simultaneous equations for analysing the problem. Investigating the numerical solutions of the meshless, the BEM, the FEM, and the WFDM and comparing those with the observed results obtained using a new model constructed in our laboratory, we could find an approach to obtain some evidence to explain the fact that the distance reached by the DO-rich water was more than 300 metres in the B reservoir in spite of the very low-velocity water flow.

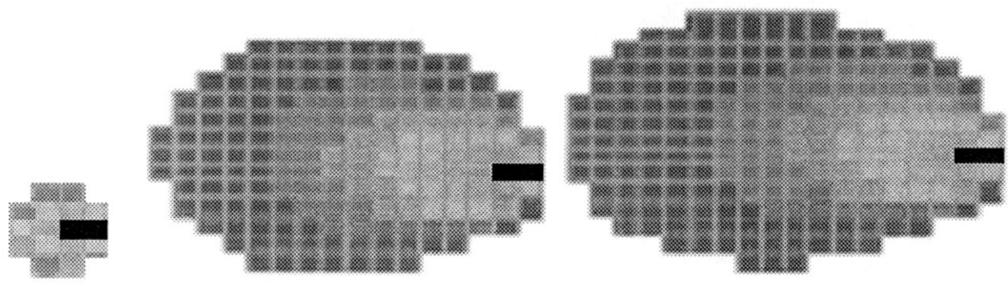

t = 2sec t = 100sec t = 200sec

Figure 7: DO-concentration distribution using WFDM (the contours were drawn when C was larger than 12mg/L).

5 Conclusion

In summary, (1) the newly investigated boundary element method could give stability and convergence to the flow analysis around the DO-supplying machine; (2) the meshless method could be used to calculate two kinds of problems; the first is the flow caused by the DO-supplying machine, and the second is the DO-concentration distribution around the DO-supplying machine. The stability and convergence of the two kinds of analysis using the meshless method seemed satisfactory; (3) field studies in a few water reservoirs led to reports of a phenomenon in which the distance reached by the DO-rich water was more than 300 metres in spite of the very low velocity of the water flow; (4) investigating the numerical solutions of the meshless method, the BEM, the FEM, and the WFDM and comparing them with the observed results obtained using a new model constructed in our laboratory, we found an approach to obtain some evidence to explain the phenomena described above.

References

[1] Kanoh, M., Nakamura, N., Kuroki T., & Sakamoto, K., Meshless analysis of flow and concentration in a water reservoir, *Proc. of the 29th World Conf. on Boundary Elements and Other Mesh Reduction Methods*, ed. C.A. Brebbia, WIT PRESS, Southampton, UK, pp. 169–178, 2007.

[2] Kanoh, M., Nakamura, N., and Kuroki T., Boundary element method for the analysis of flow and concentration in a water reservoir, *Proc. of the 28th World Conf. on Boundary Elements and Other Mesh Reduction Methods*, ed. C.A. Brebbia, WIT PRESS, Skiathos, Greece, pp. 231–240, 2006.

[3] Kanoh, M., Nakamura, N., and Kuroki T., Flow analysis in a water reservoir using a combined boundary element and weighted finite difference method, *Proc. of 27th World Conf. on Boundary Elements and Other Mesh Reduction Methods*, ed. C.A. Brebbia, WIT PRESS, Florida, USA, pp. 429–438, 2005.

Boundary element method for micropolar fluid flow in a channel

M. Zadravec, M. Hriberšek & L. Škerget
University of Maribor, Faculty of Mechanical Engineering, Slovenia

Abstract

The contribution deals with numerical simulation of micropolar fluid flow in channel, describing flow of suspensions with rigid and underformable particles with own rotation. The micropolar fluid flow theory is incorporated into the framework of a velocity-vorticity formulation of Navier-Stokes equations. The governing equations are derived in differential and integral form, resulting from the application of a Boundary Element Method (BEM). In integral transformations, the diffusion-convection fundamental solution for flow kinetics, including vorticity transport, heat transport and microrotation transport, is implemented. The Poiseuille flow test case is the benchmark case of channel flow. The results show, that the ratio between vortex viscosity coefficient and spin gradient viscosity coefficient controls the microrotation in the micro channel.
Keywords: boundary element method, micropolar fluid, poiseuille flow.

1 Introduction

In the recent years there was a progress in a micromachining technology. Opinion of a few scientists is that flows on the microscale differ from that on a macroscale, described by the Navier-Stokes equations. To clarify such assumption micropolar fluid theory is gaining interest of a lot of researchers. Micropolar fluids are a subclass of microfluids, introduced by Eringen [1]. A simple microfluid is by Eringen's definition a fluid medium whose properties and behaviour are influenced by the local motions of the material particles contained in each of its volume elements. A microfluid is isotropic viscous fluid and possesses local inertia. Because of a complex formulation for a general microfluid this class of fluids is divided into subclasses, which allow a simplified

WIT Transactions on Modelling and Simulation, Vol 47, © 2008 WIT Press
www.witpress.com, ISSN 1743-355X (on-line)
doi:10.2495/BE080041

description of the effects arising from particle micromotions. As mentioned, in micropolar fluids, a subclass of microfluids, rigid particles contained in a small volume element can rotate about the center of the volume element, which is described by the micro-rotation vector (Eringen [1,2]). This local rotation of the particles is independent of the mean fluid flow and its local vorticity field. Lukaszewicz [3] presented in his book mathematical aspects of the micropolar fluid flow theory. From this theory it is also expected to successfully describe non-Newtonian behaviour of certain fluids, such as liquid crystals, ferro liquids, colloidal fluids, liquids with polymer additives, animal blood carrying deformable particles (platelets), clouds with smoke, suspensions, slurries and liquid crystals. This theory gain attention of researchers in recent years also in other fields of fluid flows.

Papautsky *et al* [4] described microchannel fluid flow behaviour with numerical model based on micropolar fluid theory and experimentally verified the model. Results showed that micropolar fluid theory present better agreement with experiment than the use of classical Navier-Stokes theory. In the paper of Pietal [5] it is evident that applicability of the micropolar fluid theory in microchannels depends on the geometrical dimension of the flow field.

Among different approximation methods for solving problems of fluid flow the Boundary Element Method (BEM) is a relatively new method with some interesting features, described in Škerget *et al* [6,7], Hriberšek and Škerget [8]. Here, we will focus on the development of BEM for velocity-vorticity formulation of Navier-Stokes equations presented by Škerget *et al* [6] and show how to incorporate the micropolar fluid theory into the BEM framework.

2 Mathematical formulation

For description of compressible viscous fluid flow we use conservation laws for mass, momentum and energy with appropriate rheological models and equations of state. In the case of isothermal fluids which behaviour can be described by micropolar fluid flow theory Eringen [1] presented modified equation of conservation laws for mass (1), momentum (2) and microrotation (3):

$$\frac{\partial \rho}{\partial t} + \vec{\nabla}(\rho \vec{v}) = 0 \tag{1}$$

$$\begin{aligned} \rho \frac{D\vec{v}}{Dt} = -\vec{\nabla} p + (\lambda_v + 2\mu_v + k_v)\vec{\nabla}\vec{\nabla} \cdot \vec{v} - \\ -(\mu_v + k_v)\vec{\nabla} \times \vec{\nabla} \times \vec{v} + k_v \vec{\nabla} \times \vec{N} + \rho \vec{f} \end{aligned} \tag{2}$$

$$\rho j \frac{D\vec{N}}{Dt} = (\alpha_v + \beta_v + \gamma_v)\vec{\nabla}\vec{\nabla} \cdot \vec{N} - \gamma_v \vec{\nabla} \times \vec{\nabla} \times \vec{N} + k_v \vec{\nabla} \times \vec{v} - 2k_v \vec{N} + \rho \vec{l} \tag{3}$$

Differential operator $D(\cdot)/Dt = \partial(\cdot)/\partial t + v_k \partial(\cdot)/\partial x_k$ represents the Stokes material derivative. In the next step we assume that fluid mass density ρ and all micropolar fluid properties as second order viscosity coefficient λ_v, dynamic viscosity μ_v, vortex viscosity coefficient k_v, viscosity gradient coefficients $\alpha_v, \beta_v, \gamma_v$ and microinertia j, are constant parameters. We also consider zero couples $\vec{f}$ and $\vec{l}$. Therefore we can rewrite equations (1) - (3) for the assumption that micropolar fluid flow will be viscous, incompressible, steady state and laminar also using vector algebra $\left(\vec{\nabla} \times \vec{\nabla} \times \vec{F} = \vec{\nabla}(\vec{\nabla} \cdot \vec{F}) - \Delta\vec{F}\right)$. Considering the mass conservation and accounting for the solenoidality of microrotation field the mass, momentum and microrotation conservation equations are further simplified to:

$$\vec{\nabla} \cdot \vec{v} = 0 \tag{4}$$

$$\rho \frac{D\vec{v}}{Dt} = -\vec{\nabla} p + (\mu_v + k_v)\Delta\vec{v} + k_v \vec{\nabla} \times \vec{N} \tag{5}$$

$$\rho j \frac{D\vec{N}}{Dt} = \gamma_v \Delta\vec{N} + k_v \vec{\nabla} \times \vec{v} - 2k_v \vec{N} \tag{6}$$

To incorporate micropolar fluid flow theory into the framework of velocity-vorticity formulation of Navier-Stokes equations and to apply the BEM approximation method, we must first split the dynamics of the flow into its kinematic and kinetic part. This is done by the use of derived vector vorticity field function $\vec{\omega}$, obtained as a curl of the compatibility velocity field $\vec{\omega} = \vec{\nabla} \times \vec{v}$, which is solenoidal vector function by the definition $\vec{\nabla} \cdot \left(\vec{\nabla} \times \vec{v}\right) = 0$. By applying the curl operator to vorticity and using the mass conservation equation (5) for the incompressible fluid flow we get elliptic Poisson equation for the velocity vector, Skerget *et al* [6]:

$$\Delta\vec{v} + \vec{\nabla} \times \vec{\omega} = 0 \tag{7}$$

or in tensor notation form:

$$\frac{\partial^2 v_i}{\partial x_j \partial x_j} + e_{ijk} \frac{\partial \omega_k}{\partial x_j} = 0 \tag{8}$$

Equation (8) represents kinematic part of the fluid flow where for known vorticity field, the corresponding velocity field can be determined.

To compute the kinetic part of the flow we apply the curl operator to the momentum conservation equation (5) and considering that $\vec{\nabla} \cdot \vec{\omega} = 0$, $\vec{\nabla} \cdot \vec{v} = 0$ and $\vec{\nabla} \cdot \vec{N} = 0$ due to the vorticity and microrotation definition and mass conservation equation, it follows:

$$\rho \frac{D\vec{\omega}}{Dt} = (\mu_v + k_v)\Delta\vec{\omega} + \left(\vec{\omega} \cdot \vec{\nabla}\right)\vec{v} - k_v \Delta\vec{N} \tag{9}$$

This equation shows that the rate of change of the vorticity field is due to viscous diffusion, vortex stretching and twisting, and microrotation.
For the case of two-dimensional plane flow and accounting for all previous assumption the final form of equations for kinematic and kinetic part is expressed in Cartesian tensor notation form as:

$$\frac{\partial^2 v_i}{\partial x_j \partial x_j} + e_{ij}\frac{\partial \omega}{\partial x_j} = 0 \tag{10}$$

$$\rho\frac{D\omega}{Dt} = (\mu_v + k_v)\frac{\partial^2 \omega}{\partial x_j \partial x_j} - k_v\frac{\partial^2 N}{\partial x_j \partial x_j} \tag{11}$$

$$\rho j\frac{DN}{Dt} = \gamma_v\frac{\partial^2 N}{\partial x_j \partial x_j} + k_v e_{ij}\frac{\partial v_i}{\partial x_j} - 2k_v N \tag{12}$$

If we assume that $k_v = 0$ the equation for vorticity (11) and microrotation (12) are uncoupled, therefore the flow is independent of the microrotation, and the governing equations now resume the form of the classical Navier-Stokes equations.

When we use the velocity-vorticity formulation the pressure term is eliminated from momentum conservation equation (5). In order to compute pressure distribution we derive the equation for pressure, which is obtained from equation (5) with considering mass conservation equation (7):

$$\vec{\nabla}p = -\rho\frac{D\vec{v}}{Dt} - (\mu_v + k_v)\vec{\nabla}\times\vec{\omega} + k_v\vec{\nabla}\times\vec{N} = \vec{f}_{pm} \tag{13}$$

To derive the pressure equation, depending on known velocity, vorticity and microrotation field, the divergence of equation (13) should be calculated, resulting in elliptic Poisson pressure equation:

$$\Delta p - \vec{\nabla}\cdot\vec{f}_{pm} = 0 \tag{14}$$

The advantage of the Boundary Domain Integral Method, originating from Boundary Element Method, lies in the application of Green's fundamental solutions as particular weighting functions. Different conservation models can be written with an appropriate selection of a linear differential operator $L[u]$ in the following general form:

$$L[u] + b = 0 \tag{15}$$

where the operator $L[u]$ can be either elliptic or parabolic, $u(r_j, t)$ is an arbitrary field function, and the nonhomogenous term $b(r_j, t)$ is applied for non-linear transport effects or pseudo body forces.

The integral representation of flow kinematics (10) is derived from the elliptic Poisson partial differential equation, Škerget *et al.* [6], resulting in:

$$c(\xi)v_i(\xi) + \int_{\Gamma} v_i \frac{\partial u^*}{\partial n} d\Gamma = e_{ij} \int_{\Gamma} v_j \frac{\partial u^*}{\partial n_t} d\Gamma - e_{ij} \int_{\Omega} \omega \frac{\partial u^*}{\partial x_j} d\Omega \tag{16}$$

where u^* is elliptic fundamental solution.

To apply integral representation based on diffusion-convection fundamental solution for flow vorticity (11) and microrotation (12), nonhomogenous velocity field $\vec{v}(r)$ must be decomposed into the constant $\vec{v}_o$ and a variable or perturbated part $\tilde{\vec{v}}(r)$. A general form of the diffusion-convective equation with first order chemical reaction term is therefore:

$$\frac{\partial u}{\partial t} + \frac{\partial u v_{oj}}{\partial x_j} = a_o \frac{\partial^2 u}{\partial x_j \partial x_j} - \kappa u - \frac{\partial \tilde{v}_j u}{\partial x_j} + \frac{I_o}{c_o} \tag{17}$$

Where a_o and c_o are constant transport material properties, β is the reaction rate constant, while I_o stands for known source term. For accumulation term $\partial u/\partial t$ the first order Euler scheme discretisation is used $\partial u/\partial t \approx (u_F - u_{F-1})/\Delta t$, where the subscript F is indicating variable u in current time step and subscript $F-1$ in previous time step and Δt is the time step.
With use of equation (15) the equation (17) can be stated as:

$$L[u] + b = a_o \frac{\partial^2 u}{\partial x_j \partial x_j} - \kappa u - v_{oj} \frac{\partial u}{\partial x_j} + b \tag{18}$$

where $L[\cdot]$ is linear differential operator, and b stands for pseudo body force term. The parameter κ is defined as a sum of time increment parameter from accumulation term discretisation and χ which accounts other reaction terms:

$$\kappa = \frac{1}{\Delta t} + \chi \tag{19}$$

If we suppose that we know the fundamental solution $u^*(\xi, s)$ satisfying the equation $L^*[u^*] + \delta(\xi, s) = 0$, where $L^*[\cdot]$ denotes the adjoint operator to $L[\cdot]$ we can write:

$$a_o \frac{\partial^2 u^*(\xi,s)}{\partial x_j(s) \partial x_j(s)} - \kappa u^*(\xi,s) - v_{oj} \frac{\partial u^*(\xi,s)}{\partial x_j(s)} + \delta(\xi,s) = 0 \tag{20}$$

The fundamental solution u^* and its normal derivative is expressed as (Skerget *et al* [6]):

$$u^* = a_o \frac{1}{2\pi a_o} K_o(\zeta r) exp\left(\frac{v_{oj} r_j}{2a_o}\right) \tag{21}$$

$$\frac{\partial u^*}{\partial x_j} n_j = \frac{n_j}{2\pi r^2 a_o}\left[\zeta r K_1(\zeta r) r_j - \frac{r^2}{2a_o} K_o(\zeta r) v_{oj}\right] exp\left(\frac{v_{oj} r_j}{2a_o}\right) \tag{22}$$

where the parameter ζ is defined as:

$$\zeta^2 = \left(\frac{v_o}{2a_o}\right)^2 + \frac{\kappa}{a_o} = \left(\frac{v_o}{2a_o}\right)^2 + \frac{1}{a_o \Delta t} + \frac{\chi}{a_o} \tag{23}$$

K_o and K_1 are the modified Bessel functions of the second kind, $r_j(\xi, s)$ is the vector from the source point ξ to the reference point s, while r is its magnitude $r = |r_j|$ and $v_o^2 = v_{oj} v_{oj}$.

By applying Green's theorem for scalar field function to equation (18) and applying the Gauss theorem to the domain integral of pseudo body source term b which includes the convection for the perturbated velocity field only, source term and initial conditions from accumulation term discretisation, we can write:

$$c(\xi)u(\xi) + a_o \int_\Gamma u \frac{\partial u^*}{\partial n} d\Gamma$$

$$= a_o \int_\Gamma \frac{\partial u}{\partial n} u^* d\Gamma - \int_\Gamma u v_{on} u^* d\Gamma + \int_\Omega \tilde{v}_j u \frac{\partial u^*}{\partial x_j} d\Omega \tag{24}$$

$$+ \frac{1}{c_o} \int_\Omega I_o u^* d\Omega + \frac{1}{\Delta t} \int_\Omega u_{F-1} u^* d\Omega$$

Equation (24) represents integral equation for kinetic of the property u. The meanings of a_o, c_o and I_o for each u are listed in table 1.

Table 1: Generalised representation of kinetic equations in integral form.

Equation	u	a_o	χ	c_o	I_o
(11)	ω	$\frac{\mu_v + k_v}{\rho}$	0	$\frac{\rho}{k_v}$	$\frac{\partial^2 N}{\partial x_j \partial x_j}$
(12)	N	$\frac{\gamma_v}{\rho j}$	$\frac{2k_v}{\rho j}$	$\frac{\rho j}{k_v}$	ω

3 Results

For the test case the Poiseuille flow in the channel was chosen (Fig. 1). The mesh has 50 elements in the x axis and 40 elements in y axis. The mesh was refined in the vicinity of solid walls with refinement factor 4. On the inlet the uniform velocity was prescribed ($v_{x=0}=1 m/s$) normal to the inlet boundary.

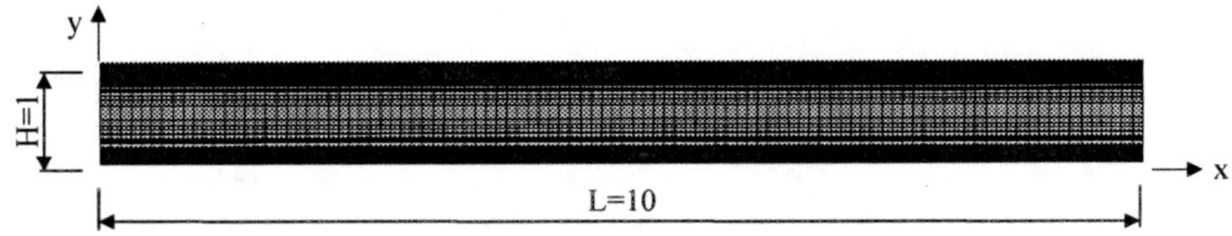

Figure 1: Geometry and mesh of the problem.

A general type of the boundary condition for the microrotation on the no slip walls for the 2D planar problem was proposed by Rees and Bassom [9]:

$$N_{wall} = -n\frac{\partial v_x}{\partial y} \qquad 0 \leq n \leq 1 \tag{25}$$

where $n = 1/2$ indicates weakly concentrated suspensions (Jenna and Mathur [10]), $n = 1$ indicates turbulent shear flows (Peddieson [11]) and $n = 0$ is called strong concentrated suspensions (Guram and Smith [12]). Our calculation was performed for the $n = 0$ and therefore microrotation was equal zero at the rigid boundaries. The initial velocity and microrotation fields were set to zero. Microinertia was defined as a square of the length scale ($j = L^2$), as stated by the Rees and Bassom [9]. Viscosity gradient coefficient γ_v or the spin gradient viscosity was taken in the form proposed by the Ahmadi [13] as:

$$\gamma_v = \left(\mu + \frac{k_v}{2}\right)j \tag{26}$$

The calculation was performed for different material parameters $K=0$, 0.1, 0.5, 10 and 100, representing the ratio between vortex viscosity and dynamic viscosity. The case $K=0$ was achieved by setting vortex viscosity to be zero. This case is presenting calculation of the micropolar fluid in which the microrotation does not affect the flow (as in the case of classical Navier-Stokes equation). On fig. 2 the microrotation profiles at the outlet are presented. An additional coefficient λ, which is a combination of vortex viscosity k_v and spin gradient viscosity coefficient γ_v, is controlling the microrotation. When the parameter k_v is zero also the coefficient λ is zero. We can see that in the case $k_v = 0$ also the coefficient λ is zero and the microrotation is zero and that increasing of coefficient λ is causing increase of microrotation. Results in fig 2 are consist with results of Eringen [1]. The pressure difference in dependence of the material parameter is shown in fig. 3.

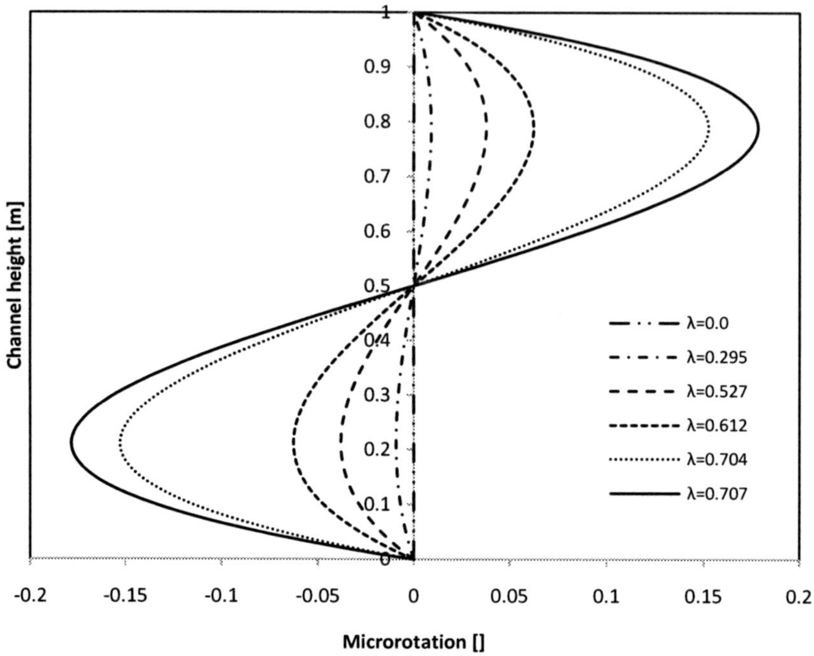

Figure 2: Microrotation profiles.

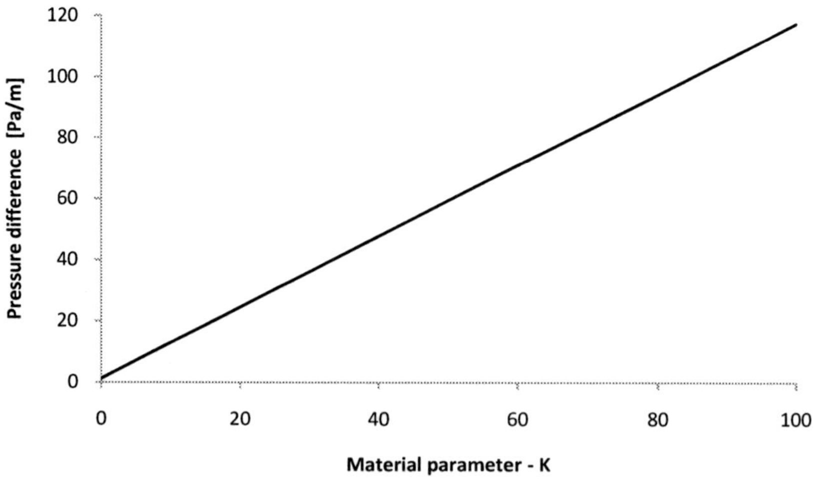

Figure 3: Pressure difference.

4 Concluding remarks

The paper presented the derivation of differential as well as integral forms of equations in the framework of velocity-vorticity formulation for numerical simulation of fluid flow with micropolar fluid theory. The derivation showed that the derived equations include several additional terms, compared with the classical set of Navier-Stokes equations. The physical model was incorporated into the BEM numerical code. The derived numerical algorithm was used to analyze the effect of different parameters of micropolar fluid flow on flow in a channel. Computational results were also compared with results of other authors and comparison shows good agreement.

References

[1] Eringen, A.C., Theory of Micropolar Fluids. *J Math Mech,* **16**, pp. 1-18, 1966.

[2] Eringen, A.C., *Microcontinuum Field Theories: II.* Fluent media. New York: Springer Verlag, 2001.

[3] Lukaszewicz, G., *Micropolar Fluids: Theory and Application.* Boston: Birkhäuser, 1999.

[4] Papautsky, I., Brazzle, J., Ameel, T., & Frazier, A.B., Laminar fluid behaviour in microchannels using micropolar fluid theory. *Sensors and actuators*, **73(1-2)**, pp. 101-108, 1999.

[5] Pietal, A.K., Microchannels flow modelling with the micropolar fluid theory. *Bulletin of the Polish Academy of Sciences*, **52(3)**, pp. 209-214, 2004.

[6] Škerget, L., Hriberšek, M. & Kuhn, G., Computational fluid dynamics by boundary-domain integral method. *Int J Numer Met Engng*, **46**, pp. 1291-1311, 1999.

[7] Škerget, L., Hriberšek, M. & Žunič, Z., Natural convection flows in complex cavities by BEM. *Int J Numer Methods Heat Fluid Flow*, **13(5/6)**, pp. 720-736, 2003.

[8] Hriberšek, M. & Škerget, L., Boundary domain integral method for high Reynolds viscous fluid flows in complex planar geometries. *Comput Methods Appl Mech Engrg*, **194**, pp. 4196-4220, 2005.

[9] Rees, D.A.S. & Bassom, A.P., The Blasius boundary-layer flow of a micropolar fluid. *Int J Engng Sci*, **34**, pp. 113-124, 1996.

[10] Jena, S.K. & Mathur, M.N., Similarity solutions for laminar free convection flow of a hermomicropolar fluid past a nonisothermal flat plate. *Int J Engrg Sci*, **19**, pp. 1431–1439, 1981.

[11] Peddieson, J., An application of the micropolar fluid model to the calculation of turbulent shear flow. *Int J Engrg Sci*, **10**, pp. 23–32, 1972.

[12] Guram, G.S. & Smith, C., Stagnation flows of micropolar fluids with strong and weak interactions. *Comput Math Appl*, **6**, pp. 213–233, 1980.

[13] Ahmadi, G., Self-similar solution of incompressible micropolar boundary layer flow over a semi-infinite flat plate. *Int J Engng Sci*, **14**, pp. 639-646, 1976.

A multipole based Boundary Element Method for moving boundary problems in axisymmetric geometry

J. Singh[1], A. Glière[1] & J.-L. Achard[1,2]
[1]*CEA-LETI MINATEC, France*
[2]*LEGI, Microfluidics, Interfaces & Particles Team, France*

Abstract

The Boundary Element Method (BEM) has been an effective method for modelling a number of particular problems which may be described by potential flows or, in a quite different vein, by creeping motions. However when it comes to moving boundary flows the BEM requires a significant amount of computational time. In the context of accelerating the BEM, the Fast Multipole Method (FMM) has been used for two and three-dimensional problems but they are not yet reported for axisymmetric ones. Our purpose is to present a multipole based boundary element method (MM-BEM) for moving boundary problems in the axisymmetric case. The proposed method takes advantage of grouping the ring sources in the axisymmetric domain, in order to reduce the amount of direct computations. Direct computations are only performed when the ring sources are located close to the evaluation point. Here, MM-BEM is implemented to simulate the impact of a drop onto a liquid surface, modelled assuming potential flow. It is shown that the proposed method provides the same results as the conventional BEM.

Keywords: multipole expansion, multipole coefficient, Boundary Element Method, potential flow, axisymmetric problem, moving boundary.

1 Introduction

A handful of practical applications characterized by moving boundaries can be modelled in axisymmetric environment, specifically the applications involving drops and bubbles [1–3]. The objective is to keep track of the moving boundary at different instants of time. The interest of the axisymmetric formulation lies in

www.witpress.com, ISSN 1743-355X (on-line)
doi:10.2495/BE080051

the fact that, while keeping the whole three-dimensional physical aspect of the problem, the solution is carried out in a two-dimensional domain. Thus, the implementation becomes easier and less computational resources are required.

The BEM has been an efficient technique for tracking moving boundaries, especially when dealing with potential and viscous fluid flow assumptions [4]. The method has tremendous advantage that the boundary condition is treated neatly and resolved very accurately. However, simulating highly distorted interfaces require distributing a large number of nodes, in which case the BEM requires a significant amount of computational time. Moreover, if the problem is solved at several time sequences the aforesaid issue turns demanding. The iterative solver has particular advantage for moving boundary problems, since the solution of the previous time step provides a good initial guess for the following step. However, the free surfaces are Dirichlet type boundary, for which the integral equation is Fredholm integral equation of the first kind, which is not sufficiently well conditioned. The numerical solution requires a large number of iterations.

In the context of accelerating the BEM, the fast iterative solver techniques, like FMM, tackles efficiently the issues of time and memory in two and three-dimensional problems. However, FMM is not applicable for axisymmetric problems, due to the presence of ring sources.

In this work we implement a multipole method coupled with BEM (MM-BEM) for moving boundary flow in the extensively studied application of drop splash, wherein a droplet is subjected to impinge on a liquid free surface. The application represents an interesting case of highly distorted moving interface and is modelled by assuming the potential flow of a fluid in an axisymmetric environment. The multipole expansion is first used to separate the ring sources from the evaluation point, which in turn enables to make a grouping structure for the ring sources. The contribution of a group is then calculated at the axis of symmetry in terms of multipole coefficients and this information is efficiently used at the evaluation point. In this manner we obtain a fast matrix vector multiplication which is coupled with an iterative solver to accelerate the solution procedure.

Outline of the paper is as follows. In the second and third sections we present the gradient and the multipole formulation of the Dirichlet problem and the general features of the MM-BEM technique. The application to liquid drop impact is presented in the fourth section. The two last sections consist of numerical results and a brief conclusion.

2 Axisymmetric gradient formulation

In the context of the potential flow theory, Dirichlet boundary conditions are applied to moving free surfaces. The integral equation formulation for the Dirichlet problem results in a Fredholm integral equation of the first kind, which is not well conditioned in terms of numerical solution. However, this Fredholm integral equation can be converted to a second kind equation for the boundary distribution of the normal derivative, which is better conditioned. The gradient formulation for the Laplace equation is well established [3], and, for completeness, we present it here for the axisymmetric case. Starting with the integral equation for the three dimensions Laplace equation defined over the surface S which bounds the domain of interest [4]

$$c(\mathbf{x})\varphi(\mathbf{x}) = \int_S \frac{\partial \varphi}{\partial \mathrm{n}'}(\mathbf{x}')G(\mathbf{x},\mathbf{x}')dS' - \int_S \varphi(\mathbf{x}')\frac{\partial G}{\partial \mathrm{n}'}(\mathbf{x},\mathbf{x}')dS', \quad (1)$$

where φ is the velocity potential, $G(\mathbf{x},\mathbf{x}')=1/4\pi \mathrm{r}$ is the free space Green's function and $\mathrm{r} = |\mathbf{x} - \mathbf{x}'|$ is the distance between the evaluation point $\mathbf{x}$ and the source point $\mathbf{x}'$. Moreover, $\partial\varphi/\partial\mathrm{n} = \mathbf{n}.\nabla\varphi$ denotes the normal derivative with $\mathbf{n}$ the unit outward normal vector on S, and $c(\mathbf{x})$ depends on the local geometry of S at $\mathbf{x}$ ($c=1/2$ if S is regular at $\mathbf{x}$). From eqn (1) an integral equation for the normal derivative of the potential at the evaluation point can be derived as

$$c(\mathbf{x})\frac{\partial \varphi}{\partial \mathrm{n}}(\mathbf{x}) = \int_S \frac{\partial \varphi}{\partial \mathrm{n}'}(\mathbf{x}')\left[\mathbf{n}.\nabla G(\mathbf{x},\mathbf{x}')\right]dS' - \int_S \varphi(\mathbf{x}')\left[\mathbf{n}.\nabla\left(\frac{\partial G}{\partial \mathrm{n}'}(\mathbf{x},\mathbf{x}')\right)\right]dS'. \quad (2)$$

The first and second surface integrals in eqn (2) express boundary distribution of point sources and point sources dipoles respectively. This equation is hypersingular due to the kernel in the second integral and can get regularized by following equivalence between doublet and vortex distributions (see e.g., reference [5])

$$\int \varphi(\mathbf{x}')\left[\mathbf{n}.\nabla\left(\frac{\partial G}{\partial \mathrm{n}'}(\mathbf{x},\mathbf{x}')\right)\right]dS' = \int \mathbf{n}.\left[(\nabla'\varphi\times\mathbf{n}')\times\nabla' G(\mathbf{x},\mathbf{x}')\right]dS'. \quad (3)$$

Let the z-axis in cylindrical polar coordinate system (ρ,ϕ,z) coincide with the axis of revolution of the domain of interest. We assume that the velocity potential is rotationally symmetric about the z-axis. The azimuthal component of the outer normal vanishes over the boundary S. The velocity potential and its gradient depend only on ρ and z. The meridian plane where the evaluation point $\mathbf{x}$ lies is arbitrary. For sake of simplicity it corresponds here to $\phi=0$. The differential area dS is expressed in the form $\rho\, d\phi\, d\Gamma$, where $d\Gamma$ is the differential arc length along the contour Γ of the surface in a meridian plane. An important consequence of axial symmetry is to simplify the integrand in the right hand side of eqn (3), since

$$\left[(\nabla'\varphi\times\mathbf{n}')\times\nabla' G\right] = -\frac{\partial \varphi}{\partial \mathrm{s}'}(\mathbf{x}')\left[\mathbf{i}_{\phi'}\times\nabla' G\right], \quad (4)$$

with $\partial\varphi/\partial\mathrm{s}'$ being the tangential component of the velocity, acting here as a vortex strength and $\mathbf{i}_{\phi'}$ is the unit vector on S at $\mathbf{x}'$ perpendicular to the meridian plane. The sense of $\mathbf{s}$, the unit tangent vector, is such that $(\mathbf{s},\mathbf{n},\mathbf{i}_{\phi})$ is right-handed in that order. Taking into account all the notations and axisymmetric simplifications, we recast eqn (2) into the form

$$\begin{aligned} c(\mathbf{x})\frac{\partial \varphi}{\partial \mathrm{n}}(\mathbf{x}) = {} & \int_\Gamma \frac{\partial \varphi}{\partial \mathrm{n}'}(\mathbf{x}')\left(\mathrm{n}_\rho\frac{\partial G^{AX}}{\partial \rho} + \mathrm{n}_z\frac{\partial G^{AX}}{\partial z}\right)\rho(\mathbf{x}')d\Gamma' \\ & - \int_\Gamma \frac{\partial \varphi}{\partial \mathrm{s}'}(\mathbf{x}')\left(\mathrm{n}_\rho\frac{\partial \widetilde{G}^{AX}}{\partial z} + \mathrm{n}_z\frac{\partial G^{AX}}{\partial \rho'}\right)\rho(\mathbf{x}')d\Gamma', \end{aligned} \quad (5)$$

where

$$G^{AX} = \frac{1}{4\pi}\int_0^{2\pi}\frac{d\phi'}{|\mathbf{x}-\mathbf{x}'|} \text{ and } \tilde{G}^{AX} = \frac{1}{4\pi}\int_0^{2\pi}\frac{\cos\phi'}{|\mathbf{x}-\mathbf{x}'|}d\phi' \tag{6}$$

are free space axisymmetric Green's function. In the conventional BEM formulation, explicit relations for G^{AX} and $\tilde{G}^{AX}$ are expressed in terms of complete elliptic integrals of the first and second kinds. We rather adopt a multipole expansion based grouping technique.

3 Axisymmetric multipole formulation

In this section we describe the multipole techniques for the solution of eqn (5) in order to accelerate the matrix vector multiplication for iterative solver. The evaluation and source points are first separated and then an azimuthal integration is carried out to obtain the axial multipole moments. We begin with the expansion of the Green's function associated with the three dimension Laplace equation in spherical coordinates system (r, θ, ϕ) [6]

$$\frac{1}{4\pi}\frac{1}{|\mathbf{x}-\mathbf{x}'|} = \frac{1}{4\pi}\sum_{n=0}^{\infty}\sum_{m=0}^{n}\varepsilon_m\frac{(n-m)!}{(n+m)!}\frac{r'^n}{r^{n+1}}P_n^m(\cos\theta)P_n^m(\cos\theta')\cos m(\phi-\phi'), \tag{7}$$

where ε_m is the Neumann factor, ε_0=1 and ε_i=2 (i=1,2,etc.) and P_n^m are the associated Legendre functions. To obtain the series expansion of the free space axisymmetric Green function G^{AX} we integrate the above series in the azimuthal direction, keeping the origin on the axis of symmetry

$$G^{AX} = \frac{1}{4\pi}\sum_{n=0}^{\infty}\sum_{m=0}^{n}\varepsilon_m\frac{(n-m)!}{(n+m)!}\frac{r'^n}{r^{n+1}}P_n^m(\cos\theta)P_n^m(\cos\theta')\int_0^{2\pi}\cos m(\phi-\phi')d\phi'. \tag{8}$$

Since the evaluation point lies in the meridian plane, ϕ =0 and

$$\int_0^{2\pi}\cos m(\phi-\phi')d\phi' = \begin{cases} 0, & m\neq 0 \\ 2\pi, & m\to 0 \end{cases}. \tag{9}$$

Therefore, only the m=0 terms survive and we obtain

$$G^{AX} = \frac{1}{2}\sum_{n=0}^{\infty}\frac{r'^n}{r^{n+1}}P_n(\cos\theta)P_n(\cos\theta'). \tag{10}$$

The multipole coefficients for the first integral in eqn (5) are obtained by substituting eqn (10), written as

$$\begin{aligned}&\int_\Gamma\frac{\partial\varphi}{\partial \mathrm{n}'}(\mathbf{x}')\left(\mathrm{n}_\rho\frac{\partial G^{AX}}{\partial\rho}+\mathrm{n}_z\frac{\partial G^{AX}}{\partial z}\right)\rho(\mathbf{x}')d\Gamma' \\ &=\sum_{n=0}^{\infty}\left(\mathrm{n}_\rho\frac{\partial}{\partial\rho}\left(\frac{P_n(\cos\theta)}{r^{n+1}}\right)+\mathrm{n}_z\frac{\partial}{\partial z}\left(\frac{P_n(\cos\theta)}{r^{n+1}}\right)\right)M_n(O)\end{aligned} \tag{11}$$

with $M_n(O)$ being the multipole coefficients at point O due to the ring sources in Γ, given by

$$M_n(O) = \int_\Gamma\frac{\partial\varphi}{\partial \mathrm{n}'}(\mathbf{x}')r'^n P_n(\cos\theta')\rho(\mathbf{x}')d\Gamma'. \tag{12}$$

The series expansion of the second free space axisymmetric Green's function $\widetilde{G}^{AX}$ is obtained as follows:

$$\widetilde{G}^{AX} = \frac{1}{4\pi}\sum_{n=0}^{\infty}\sum_{m=0}^{n}\varepsilon_m \frac{(n-m)!}{(n+m)!}\frac{r'^n}{r^{n+1}} P_n^m(\cos\theta)P_n^m(\cos\theta')\int_0^{2\pi}\cos\phi'\cos m(\phi-\phi')d\phi'. \tag{13}$$

Integrating the above series in the azimuthal direction, while keeping the origin on the axis of symmetry and the evaluation point in the meridian plane, we get

$$\int_0^{2\pi}\cos\phi'\cos m(\phi-\phi')d\phi' = \begin{cases} 0, & m \neq 1 \\ \pi, & m \to 1 \end{cases}. \tag{14}$$

Only the terms m=1 survive and the eqn (13) becomes

$$\widetilde{G}^{AX} = \frac{1}{2}\sum_{n=0}^{\infty}\frac{(n-1)!}{(n+1)!}\frac{r'^n}{r^{n+1}} P_n^1(\cos\theta)P_n^1(\cos\theta'). \tag{15}$$

The multipole coefficients for the second integral in eqn (5) are obtained by substituting eqn (15), written as

$$\begin{aligned} &\int_\Gamma \frac{\partial\varphi}{\partial s'}(\mathbf{x}')\left(\mathrm{n}_\rho \frac{\partial\widetilde{G}^{AX}}{\partial z} + \mathrm{n}_z \frac{\partial G^{AX}}{\partial\rho'}\right)\rho(\mathbf{x}')d\Gamma' \\ &= \sum_{n=0}^{\infty}\mathrm{n}_\rho \frac{\partial}{\partial z}\left(\frac{(n-1)!}{(n+1)!}\frac{P_n^1(\cos\theta)}{r^{n+1}}\right)\widetilde{M}_n(O) + \sum_{n=0}^{\infty}\mathrm{n}_z\left(\frac{P_n(\cos\theta)}{r^{n+1}}\right)\hat{M}_n(O) \end{aligned} \tag{16}$$

where the multipole coefficients are given by

$$\widetilde{M}_n(O) = \int_\Gamma \frac{\partial\varphi}{\partial s'}(\mathbf{x}')r'^n P_n^1(\cos\theta)\rho(\mathbf{x}')d\Gamma', \tag{17}$$

$$\hat{M}_n(O) = \int_\Gamma \frac{\partial\varphi}{\partial s'}(\mathbf{x}')r'^{n-1}\left[nP_n(\cos\theta)\sin\theta + P_n^1(\cos\theta)\cos\theta\right]\rho(\mathbf{x}')d\Gamma'. \tag{18}$$

From the above expansions it is clear that the ring sources and the evaluation point are separated. Therefore, a grouping structure for the ring sources can be employed. The contribution at the evaluation point due to the ring sources in a group is calculated using the multipole coefficients corresponding to the group.

The above multipole coefficients are derived by assuming that the source ring is nearer to the expansion centre O than the evaluation point. In fact in the literature they are termed as exterior multipole moments. Since the interest is towards maximizing the multipole calculation we use also the interior multipole coefficients, which can be derived on the same lines [7].

Admissibility criterion for the multipole calculations depends upon the ratio of distance of the group from the expansion centre to that of distance of the evaluation point. The distance of the group from the expansion centre is taken as the average of the distances of the elements which are contained in the group from the expansion centre. The contribution at the evaluation point in group D (Figure 1) from the ring sources contained in group A uses the exterior multipoles coefficients for group D computed at expansion centre O. In the vice-versa case, (ring sources in group D and evaluation point in group A) the contribution is calculated by interior multipoles at expansion centre O.

The description of the method based on multipole technique is as follows:

- First, the ring sources are grouped. This has been done by defining an arc length parameter along the contour, as shown in Figure 1.
- Second, the positions of the multipole expansion centres are determined, based on the geometry of the problem. For example the points O and O' are sufficiently good choice for the cylindrical geometry (Figure 1).
- Third, the calculation of the interior and the exterior multipoles for the grouped ring sources at the multipole expansion centres are performed.
- Finally, the contribution at the evaluation point due to a particular group is computed by multipole coefficients if the admissibility criterion is satisfied; otherwise the direct calculations are used for each element of the group.

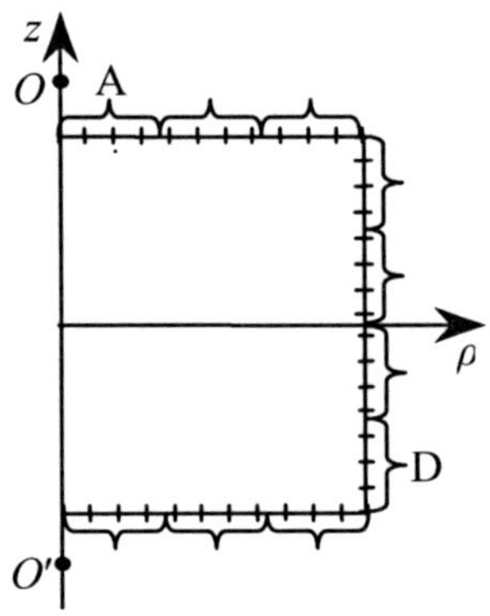

Figure 1: Grouping structure and expansion centres for a cylindrical geometry.

4 Application to drop impact

The impact of a liquid drop onto a liquid surface, which is an example of a highly distorted moving interface, is chosen as an application case for the multipole based BEM. Numerous theoretical, experimental and numerical studies have been performed during more than one century in order to better understand this phenomenon which, apart from being interesting and beautiful, is of considerable practical importance. Among other numerical techniques, the Volume Of Fluid [8], the level-set [9] and the Boundary Element Method [1] have been used.

4.1 Problem statement

An axisymmetric liquid domain, bounded in the meridian plane (ρ,z) by the liquid-gas interface Γ is considered (Figure 2). Due to the impulsive nature of the drop impact, an irrotational flow model is appropriate. A harmonic velocity potential subject to Dirichlet boundary conditions, can thus be defined. The local time derivative of the velocity potential is calculated by combining the normal momentum balance with the Laplace pressure jump to obtain the Bernoulli's equation. A dimensionless formulation is used:

$$\frac{\partial \varphi}{\partial t} = -\frac{v^2}{2} - \frac{H}{We} - \frac{z}{Fr} - \frac{2}{Re}\frac{\partial^2 \varphi}{\partial n^2} \tag{19}$$

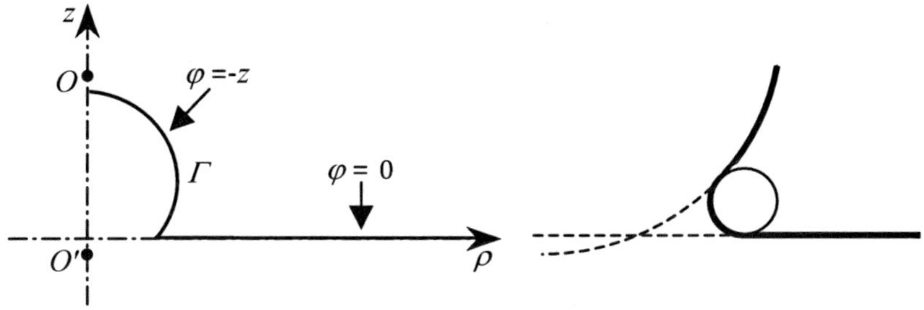

Figure 2: Initial geometry and potential. The neck region is smoothed.

In this equation v is the dimensionless velocity, H is the total curvature and Re, Fr and We are respectively the Reynolds, Froude and Weber numbers. The drop radius R is adopted as length scale and $\sqrt{\sigma/R\rho_L}$ as velocity scale, where σ and ρ_L are the surface tension and the liquid density. The ratio between the length and velocity scale gives the time scale. Viscous effects are partially taken into account as the normal viscous stress appears in eqn (19). The complete description of this model can be found in Georgescu et al. [2].

In order to avoid topological change, the falling spherical drop is initially placed in contact with the liquid surface. A negligibly small volume of the sphere is cut-out and, to avoid a singularity of the capillary pressure, the intersection of the sphere with the initially flat liquid surface is smoothed in the meridian plane by a circular element (Figure 2). The drop initially falls with a constant velocity onto an initially still liquid surface. The discontinuity of the velocity potential in the neck region is eliminated by means of the quadratic smoothing procedure described by Weiss and Yarin [10].

4.2 Solution procedure

The application represents a transient free-boundary problem that repeatedly involves two types of calculations: (a) solution of Laplace equation for normal components of the velocity, (b) updating of the potential at the forthcoming time step by using time marching scheme of Runge-Kutta and displacement of the interface.

Within the first type of calculation, the Laplace equation is solved by the Boundary Element Method using gradient formulation. First order elements are used to approximate the unknowns. The vortex strength is calculated by interpolating the given velocity potential over the interface by the cubic splines. The solution is obtained by grouping source/vortex rings using multipole techniques discussed in previous section coupled with the GMRES iterative solver without preconditioning. In order to maximize the number of multipole calculations the multipole coefficients are calculated at two different expansion centres. Two expansion centres are used and their positions are evaluated at each time step, following the deformation of the interface. The normal velocity obtained at previous time step is used as an initial guess for the iterative solver for the subsequent time step.

Within the second type of calculation, the Bernoulli's equation is solved by 4th order explicit Runge-Kutta scheme for the local time derivative of the velocity potential. The time steps are chosen according to the stability criterion described by Georgescu et al. [2]. The interface position is then updated according to the normal velocity component.

5 Results

The presented MM-BEM has been implemented for the axisymmetric static potential problems. With 18 terms in the expansion, this method provides accurate results when compared to analytical solution of Dirichlet or Neumann problems. The computational time is almost reduced by half when compared to the conventional BEM [7].

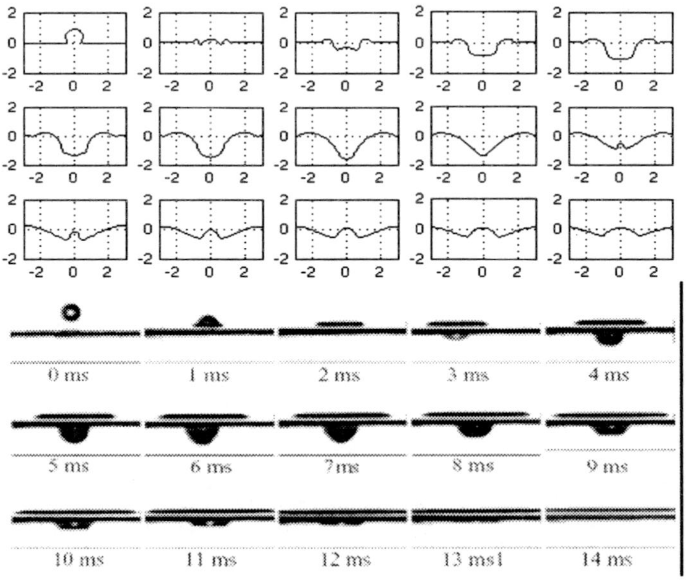

Figure 3: Comparison of numerical and experimental profiles (Fr=90, We=43) obtained by MM-BEM and Liow [11]. Bottom image reproduced with permission from Cambridge University Press.

The method is adapted for free surface flows and specifically to the chosen application of a drop impinging on a liquid surface. The number of nodes on the interface varies from 20 to 150 and the multipole series expansion is truncated after 21 terms. The comparison of the free surface profiles obtained by simulation through MM-BEM and experiments [11] at different instants are shown in Figure 3 (*Fr*=90, *We*=43). The qualitative agreement is correct.

Furthermore, a 10% of total CPU time saving is obtained when using MM-BEM instead of the conventional BEM. Simulated profiles are identical in both methods.

6 Conclusion

A multipole based BEM has been implemented in an axisymmetric environment for the simulation of moving boundary problem of drop impact onto a liquid surface. By multipole techniques the interaction between individual nodes and ring sources in far field has been replaced by the interaction between node and group of ring sources. It has been shown that results obtained through MM-BEM matches well with available experimental data.

Although the time reduction is not significant yet, there is a wide scope of improvement in MM-BEM. The tuning of the parameters, for example, the number of terms in multipole series expansion, efficient choice of multipole centre in order to increase the validity of the far field and optimizing the admissibility criterion can further lead to more CPU time saving. To the best of our knowledge, this is the first attempt to accelerate the axisymmetric moving boundary value problem using a multipole expansion based technique.

References

[1] Oguz, H.N. & Prosperetti, A., Bubble entrainment by the impact of drops on liquid surfaces. *Journal of Fluid Mechanics*, 219, pp. 143–179, 1990.

[2] Georgescu, S.C., Achard J.-L. & Canot, E., Jet drops ejection in bursting gas bubble processes. *European Journal of Mechanics B-Fluids*, 21(2), pp. 265–280, 2002.

[3] Pozrikidis, C., Three-dimensional oscillations of inviscid drop induced by surface tension. *Computer and Fluids*, 30, pp. 417–444, 2001.

[4] Brebbia, C.A., Telles, J.C.F. & Wrobel, L.C., *Boundary Element Techniques*, Springer-Verlag: Berlin and New York, 1984.

[5] Brockett, T.E., Kim, M-H. & Park, J-H., Limiting forms for surface singularity distributions when the field point is on the surface. *Journal of Engineering Mathematics*, 23, pp. 201–31, 1989.

[6] Morse, P.M. & Feshbach, H., *Methods of theoretical physics*, McGraw-Hill: New York, 1953.

[7] J. Singh, A. Glière & J.-L. Achard, Multipole accelerated BEM for axisymmetric potential problem. (in preparation).

[8] Morton, D., Rudman, M. & Liow, J. L., An investigation of the flow regimes resulting from splashing drops. *Physics of Fluids,* **12**(4), pp. 747–763, 2000.

[9] Watanabe, Y., Saruwatari, A., & Ingram, D.M., Free-surface flows under impacting droplets. *Journal of Computational Physics*, 227(4), pp. 2344–2365, 2008.

[10] Weiss, D.A. & Yarin A.L., Single drop impact onto liquid films: neck distortion, jetting, tiny bubble entrainment, and crown formation. *Journal of Fluid Mechanics*, 385, pp. 229–254, 1999.

[11] Liow, J.L., Splash formation by spherical drops. *Journal of Fluid Mechanics*, 427, pp. 73–105, 2001.

Section 2
Heat transfer

SHP solution of effect of elevated temperature on tunnel lining

P. Procházka & S. Peskova
CTU Prague, Civil Engineering, Prague, Czech Republic

Abstract

Smooth hydrodynamic particle solution seems to be a suitable method for solving problems of influence of high temperatures on the FR concrete tunnel lining and surrounding rock. This meshless method suffers from one unpleasant property: it is not easy to describe the inhomogeneous geometrical boundary conditions. In our case, the highly elevated temperature is increasing and decreasing in a time scale on the boundary. After reformulating the problem into a form obeying boundary conditions being suitable for application of the SPH method, the 2D problem can be solved. This restriction to 2D is an impact of assumption that the fire is triggered along the length of the axial direction of tunnel. This follows from the observation at concrete locations where conflagrations inside of the tunnels took part. The most extreme temperature is prevailingly considered at 1200°C to fulfill European standards. As the mechanical and physical properties change nonlinearly a set of experiments in furnaces have been conducted to improve the characteristics of the material of tunnel lining and rock. Coupled modeling is applied to conduct convergence analysis providing results from experiments and numeric method to be in compliance. Simultaneous nonlinear equations are first formulated, involving stress analysis, influence of pore pressure, change of temperature, moisture, and degree of saturation. Nonlinear mechanical properties are based on the change of mechanical parameters, as are available from experimental studies. A couple of examples will follow the theory.

Keywords: smooth particle hydrodynamics method, meshless approach, elevated temperature, tunnel lining and surrounding rock, partition of unity.

WIT Transactions on Modelling and Simulation, Vol 47, © 2008 WIT Press
www.witpress.com, ISSN 1743-355X (on-line)
doi:10.2495/BE080061

1 Introduction

The problem of fire defense of tunnel linings stems from the early beginning of tunnel construction. In an age where vehicles are now powered by explosive fuel, used both in passenger vehicles and freight transport, if an accident occurs then there can be a disastrous aftermath to the tunnel lining and sometimes also to the surrounding rock both of which can lead to the supporting system of such tunnels to become no longer serviceable. A couple of experimental studies are published on this topic, although until today experiments of this kind appear to be relatively very expensive. The experiments show the overall properties, failure strength of material, surface cracking, and other phenomena, which can be seen from the outside of the sample tested. Of greater interest to researchers and engineers is the knowledge about the process developing inside the sample, mainly the reaction of the material to sudden change of temperature on the boundary of the sample. This can be solved by some appropriate numerical methods. One of the most appropriate for this problem seems to be smoothed particle hydrodynamics (SPH), which belongs to mesh-free methods, historically developed for astrophysical applications, [1, 2]. The inherent benefit of the SPH formulation consists in transformation of partial differential equations to a system of linear algebraic using regularization. This transformation is, among others, suitable for parallel computations. Recently, SPH has grown into a successful and respected numerical tool. In particular, this method does not differ between 3D, 2D and 1D problems, as the problems defined in higher order spaces can be simulated as easy as that in 1D. An excellent review of the advantages and recent progress in SPH can be found in [3, 4]. Some problems occur when geometrical boundary conditions should be involved. Takeda *et al.* [5] proposed ghost particle method, in which some particles are located outside the domain. Heat conduction problem are solved in [6], where Taylor series expansion approximates the regularization kernels.

2 Regularization of functions and their derivatives

The concept behind SHP is based on an interpolation scheme. From mathematical calculus it is well known, [7], that for each generalized function f defined on a domain $V \subset R^n$ with boundary S there exists a positive ε and a finite cover $\{\Omega\}_{i=1}^{N} \subset V, i = 1,...,N$ (for each point $x \in V$ there is an index $i \in 1,...,N$ so that $x \in \Omega_i$) with measure of $\Omega_i < \varepsilon$ so that on Ω_i there exists function $\omega_\varepsilon^i \in C^\infty(\overline{\Omega})$, supp $\omega_\varepsilon^i \in \Omega_i$ (sometimes called cap function) which regularize the function f in such a way that f can be expressed as

$$f(\boldsymbol{x}) = \sum_{i=1}^{N} \int_{\Omega_i} f(\xi)\omega_\varepsilon^i(\boldsymbol{x} - \xi)\,\mathrm{d}\xi = f * \omega_\varepsilon^i, i = 1,...,N \,, \tag{1}$$

and the left hand side of the latter relation is called the regularization, $f * \omega_\varepsilon^i$ is the convolution. Recall some basic properties of the regularization: the volume of each cap function is unity, is equal to one. If the function f is uniform (equal to one) and $\varepsilon \to$ infinity the regularization turns to be density of the function f, for example density of probability. If $\varepsilon \to 0$ the kernel ω_ε^i turns to be the Dirac function. For each positive ε the regularization (kernel, cap function) ω_ε^i can be created infinitely differentiable (for definition of types of cap functions, see [7], for example).

Since different cap functions should be created for different Ω_i, the above definition becomes inconvenient. In order to improve this put $\omega_\varepsilon^i \equiv \omega_\varepsilon$ and the shape of Ω_i remains same for all i, the area of a circle in 2D or the volume of a sphere, for example. Now inside of the domain V select a set of points $\boldsymbol{x}_i, i = 1,\dots,N$, $\boldsymbol{x}_i$ is centered at Ω_i and a new function F is defined as

$$F(\boldsymbol{x}_i) = \sum_{i=1}^{N} \int_{\Omega_i} f(\xi)\omega_\varepsilon(\boldsymbol{x}_i - \xi)\,\mathrm{d}\xi, i = 1,\dots,N \tag{2}$$

which is formally similar to relation (1), so that it fulfils basic properties above mentioned. Since the former assumptions take place the function F cannot be expected to be equal to f any longer, but a special case: $\varepsilon \to 0$ in the sense of definition of the Dirac function.

In our case 2D problem is considered and degrees of freedom are concentrated at nodes $\boldsymbol{x}_i \in \Omega_i, i = 1,\dots,N$, Ω_i are considered as areas of the circles in which x_i is centered. In the approximation, the smoothed (regularized) function F for any physical quantity f is identified with the original function, i.e. $F \equiv f$. Moreover, the kernel ω_ε is simplifies for real calculations and the simplification is denoted as W_ε. Introducing this to (2) and setting $f_i = f(\boldsymbol{x}_i)$ gives:

$$f_i = f(\boldsymbol{x}_i) = \int_{\Omega_i} f(\xi)W_\varepsilon(\boldsymbol{x}_i - \xi)\mathrm{d}\xi \tag{3}$$

Equation (3) is the kernel representation to average functional distribution. In our next considerations additional properties of W_ε will be required:

positivity: $\quad W_\varepsilon(\boldsymbol{x}_i - \xi) \geq 0, \quad \xi \in \Omega_i$

normalization: $\quad \int_{\Omega_i} W_\varepsilon(\boldsymbol{x}_i - \xi)\mathrm{d}\xi = 1, \quad \forall\varepsilon > 0$

surface smoothness on $\partial\Omega_i$: $\quad W_\varepsilon(\boldsymbol{x}_i - \xi) = \nabla W_\varepsilon(\boldsymbol{x}_i - \xi) = \nabla\nabla W_\varepsilon(\boldsymbol{x}_i - \xi) = 0, \quad \xi \in \partial\Omega_i$

The last property follows from the fact that the order of differential equations, which are to be studied, is two, and so is the required regularity (continuity).

Using integration by parts, from the boundary conditions on $\partial\Omega_i$ it immediately follows that

$$\int_{\Omega} \nabla W_{\varepsilon}(x_i - \xi) \mathrm{d}\xi = \int_{\Omega} \nabla\nabla W_{\varepsilon}(x_i - \xi) \mathrm{d}\xi = 0 \tag{4}$$

For the sake of simplicity the approximation of the kernel W_{ε} is represented by

$$W_{\varepsilon}(x_i - \xi) = C(1 - 10r^3 + 15r^4 - 6r^5), \quad r = |\xi - x_i|/\varepsilon \quad \text{for 2D} \tag{5}$$

where $C = \dfrac{2}{h}$ for 1D problem, $C = \dfrac{7}{\pi h^2}$ for 2D problem, and $|\xi - x_i|$ is the distance between the pertinent points.

If we consider volume (area, interval) of an element $\Omega_i = \dfrac{m_i}{\rho_i}$, where m_i is the mass of the element and ρ_i is the density, using rectangular rule of evaluation of integrals yields:

$$f_i = f(x_i) = \sum_{r_{ij} \le h} \frac{m_j f_j}{\rho_j} W_{\varepsilon}(r_{ij}), r_{ij} = |x_j - x_i|, \nabla f_i = \rho_i \sum_{r_{ij} \le h} m_j \left(\frac{f_i}{\rho_i^2} + \frac{f_j}{\rho_j^2} \right) \nabla W_{\varepsilon}(r_{ij}) \tag{6}$$

3 Governing equations

According to [8], the governing equations for coupled heat-seepage-stress problem are listed as:

$$\frac{\partial \rho}{\partial t} = -\rho \frac{\partial v^{\alpha}}{\partial x^{\alpha}} \tag{7}$$

$$C\frac{\partial T}{\partial t} = \frac{\partial}{\partial x}\left(\lambda_x \frac{\partial T}{\partial x}\right) + \frac{\partial}{\partial y}\left(\lambda_y \frac{\partial T}{\partial y}\right) + \frac{\partial}{\partial z}\left(\lambda_z \frac{\partial T}{\partial z}\right) + \\ + C_w \rho_w \left(V_x \frac{\partial T}{\partial x} + V_y \frac{\partial T}{\partial y} + V_z \frac{\partial T}{\partial z} \right) \tag{8}$$

$$S\frac{\partial p}{\partial t} = \frac{\partial}{\partial x}\left(K_x \frac{\partial p}{\partial x}\right) + \frac{\partial}{\partial y}\left(K_y \frac{\partial p}{\partial y}\right) + \frac{\partial}{\partial z}\left(K_z \frac{\partial p}{\partial z}\right) + Q \tag{9}$$

$$V_x = K_x \frac{\partial p}{\partial x}, \quad V_y = K_y \frac{\partial p}{\partial y}, \quad V_z = K_z \frac{\partial p}{\partial z} \tag{10}$$

$$K_x = K_{0x} \exp(-\eta \sigma_x^{\text{eff}}), \quad K_y = K_{0y} \exp(-\eta \sigma_y^{\text{eff}}), \quad K_z = K_{0z} \exp(-\eta \sigma_z^{\text{eff}}) \tag{11}$$

where

T, p, C and λ are the temperature, the pressure, the volumetric heat capacity and the thermal conductivity of the lining and surrounding rock, respectively;

S, K are the water-supply degree and the hydraulic conductivity, respectively;

η is a constant, received from appropriate tests;

σ_x^{eff}, σ_y^{eff} and σ_z^{eff} are the effective stresses in x, y and z directions, respectively;

C_w and ρ_w are the specific heat and the density, respectively, of water.

Equation (7) describes continuity, (8) is a diffusion equation for temperature distribution, (9) is a diffusion equation for pore pressure, and momentum equation has to be added to complete the system of the coupled problem:

$$\frac{\partial v^{\alpha}}{\partial t} = -\frac{1}{\rho}\frac{\partial \sigma^{\alpha\beta}}{\partial x^{\beta}} \tag{12}$$

the moving equation for particles is

$$\frac{\partial u^{\alpha}}{\partial t} = v^{\alpha} \tag{13}$$

where $u^{\alpha} = x_{new}^{\alpha} - x_{old}^{\alpha}$ i.e. a component of the displacement is difference between new and old values of the pertinent components of coordinates.

In the sense of previous explanation the above introduced strongly nonlinear equations can be solved using iteration at each time step. The way of development of temperature due to fire extent at the surface of tunnel lining is described from experiments. The only problem in applications of SPH appears to be inhomogeneous boundary conditions. At each time step we change the elliptic equation of the type

$$Ax = b \text{ in } \Omega,\ x = \bar{x} \text{ on } \Gamma,$$

where Ω, Γ are domain and its boundary, into

$$A\xi = c, \xi = x - \bar{x}, c = b - A\bar{x} \text{ in } \Omega, \xi = 0 \text{ on } \Gamma.$$

In the case of homogeneous boundary conditions being prescribed, the problem is easily solvable. Note that the operator A can be considered linear at each time step (it actually is pseudo-linear).

4 Young's modulus

The spalling of the concrete lining during combustion in a tunnel and deterioration of the material properties is not easy to obtain from experiments. The only information, which appears to be reliable, is the distribution of Young's modulus in relation to the temperature. Also, this information is not quite precise as it depends on the position of a point in which the measurement takes part inside the lining. There are a couple of authors who tried to express the relation between the values of Young's modulus and higher temperature. Note that the temperature is considered between room temperature and 1200°C, according to European standards [9].

The test results of elastic modulus of concrete at high temperature are compared in [10]. It can be noted that, except for HPC, the elastic modulus decreases monotonically as the temperature increases, unlike the behavior of the compressive strength. Generally speaking, the elastic modulus has no or a little effect till temperature of 70°C after that it descends approximately in a linear way with the increase of the temperature. The secant modulus at the stress of $0.4f_c^T$ was assumed as the elastic modulus by Lu [11] and tri-linear expression between E_c^T and the temperature T was given as follows:

$$E_c^T = \begin{cases} (0.1-0.0015T)E_c & 20^\circ C \le T \le 300^\circ C \\ (0.87-0.00087T)E_c & 300^\circ C \le T \le 700^\circ C \\ 0.28E_c & T > 700^\circ C \end{cases} \quad (14)$$

where E_c^T and E_c are the elastic modulus of concrete at elevated and room temperature, respectively.

Xiao and König [10], indicated that the elastic modulus of HPC degraded much more obviously than that of NSC below 400°C, while above 400°C its degradation was almost the same as that of NSC. It was also reported in [12, 13] that the type of aggregate would influence the modulus of elasticity distinctively and E_c^T could decrease with the increase of water-cement ratio as well. The elastic modulus at elevated temperature drops much more greatly for concrete cured in water than that of cured in the air. The deterioration of elastic modulus seems to have nothing to do with the heating-cooling cycles but is related to the maximum elevated temperature during heating [14, 15].

5 Example

The problem of the heating of a lining on the upper side of the tunnel in symmetric angle of 90° (45° on each size of the vertical axis) by a temperature of 1200°C is solved. Standard fiber reinforced concrete with basalt fibers is considered, while the surrounding rock possesses material properties of class G4, which provides:

E = 1000 MPa, E_{plast} = 800 MPa, $E_{residual}$ = 500 MPa, $\nu = 0.25$, $\nu_{plast} = 0.46$, $\varphi = 25^0$, $C = 100\,\text{MPa}$, where the former quantities describe elastic modulus, modulus at plastic state, its residual value, elastic Poisson's ratio, the same at plastic state, coefficient of internal friction, and shear strength (cohesion). For completeness the material of lining is considered as:

E = 27000 MPa, E_{plast} = 10000 MPa, $E_{residual}$ = 5000 MPa, $\nu = 0.2$, $\nu_{plast} = 0.46$, $\varphi = 35^0$, $C = 150\,\text{MPa}$. The density of the concrete is 27 kN/m^3, the density of the rock is 33 kN/m^3.

Since both geometric and loading symmetry is obvious, the structure is solved as symmetric. In Fig. 1 hypsography of horizontal stresses in a cut of a half of the domain are presented. The peak stresses are concentrated to the upper and lower part of the lining, while at the center of the lining they almost disappear. As the values of stresses are much larger in the lining then in the rock, in the cut from the domain of rock there is no change of horizontal stresses registered. In the figure the points at which the variables are calculated are marked. The maximum stress is observed on the vertical axis of symmetry and possesses the value of 10 MPa in the lining.

Figure 2 shows the hypsography of normal horizontal stresses for the distribution of temperature 1200^0 at the upper face of the lining (the same loading scheme as in the starting situation) after thirty minutes. Note that ten minute intervals are contemplated in which tests of plastic and damage states in

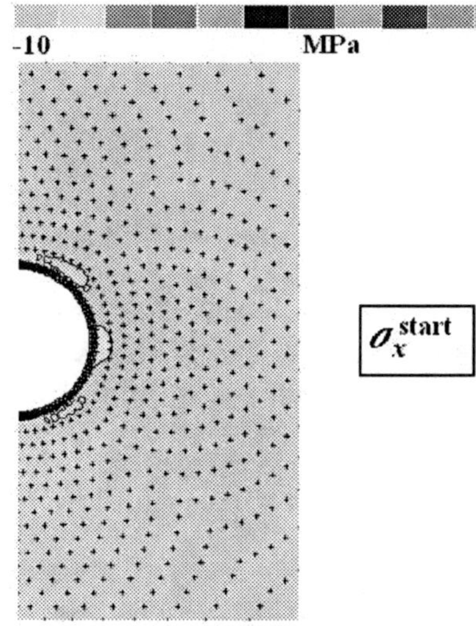

Figure 1: Horizontal stresses before fire.

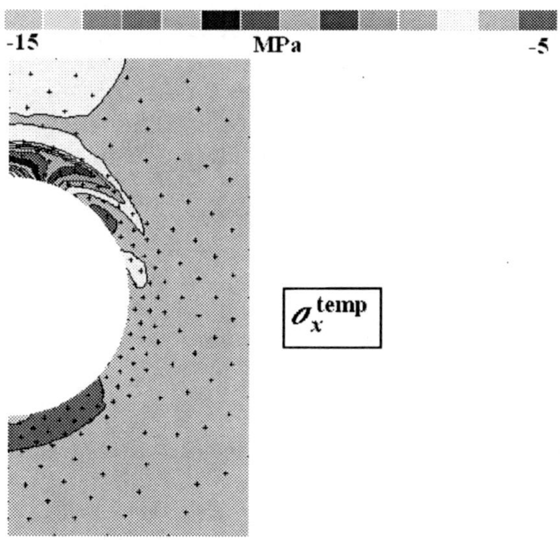

Figure 2: Horizontal stresses after half an hour combustion.

the material of both lining and rock are carried out. The latter picture is enlarged in order to see in more details the behavior of stresses σ_x in the neighborhood of the fire.

It is obvious from the figure that extreme stresses of 15 MPa are attained at the calotte (vault) of the tunnel, where the combustion is concentrated. It is worth noting that after approximately two hours of fire the concrete looses its bearing capacity and the rock takes over the loading. Rock can also be affected by the fire and damage locally. In such conditions the repair and recovery of the tunnel can be a complex task takes a long time.

6 Conclusions

In this paper the combustion of the concrete lining and the rock surrounding the tunnel is solved by the Smooth Particle Hydrodynamics Method. The results are partly compared with experiments on concrete slabs burned in a furnace. Particularly, influences of fibers from fused basalt are observed and the composite with the concrete matrix is evaluated after carrying out tests in Innsbruck University. The basalt material seems to be quite suitable for this case of endangered concrete linings. The reason is that at lower temperatures it behaves as a stiffener of the concrete matrix and overcoming the temperature of 1000°C it becomes molten and enables vapor to escape from the concrete and not cause any larger damage. Consequently, in comparison to other types of fibers, fused basalt serves in a proper way. Natural fibers are very suitable for high temperatures (they burn out and the vapor can freely dilute in the air) and stiff fibers like steel serve as a good reinforcement but are too troublesome in the case of influence of high temperatures.

The SPH method is perfectly suitable for solving such a problem. This method shows very promising time consumption of computer (it is basically low in comparing it with other numerical methods), but generally it suffers from one undesirable property: inhomogeneous geometrical boundary conditions can be respected in the calculus in a complicated way. In our case this type of boundary conditions does not play any decisive role. The boundary conditions are declared by time changing the source of heat, and the damage in the material is calculated at the time-stage.

Acknowledgements

Work on this project has been financially supported by GACR, project No. 103/06/1124. Sponsorship of CIDEAS is also acknowledged.

References

[1] Lucy, L.B. (1977). A numerical approach to testing of the fission hypothesis. *Astron. J.* **82**, 1013.

[2] Gingold, R.A. and Monoghan, J.J. (1977). Smooth particle hydrodynamics: theory and application to non-spherical stars. *Monthly Nat. R. Astron. Soc.* **181**, 375.

[3] Randle, P.W. and Libersky, L.D. (1996). Smooth particle hydrodynamics: some recent improvements and application. *Appl. Mech. Engng.* **139**, 175.
[4] Li, S. and Liu, W.K. (2002). Meshfree and particle method and their applications. *Appl. Mech. Rev.* **55**, 1.
[5] Takeda, H., Miyama, S. and Sekiya, M. (1994). Numerical simulation of viscous flow by smoothed particle hydrodynamics. *Prog. Theor. Phys.* **92**, 939.
[6] Chen, J.K., Beraun, J.E. and Carney, T.C. (1999). A corrective smooth particle method for boundary value problems in heat conduction. *Int. J. Numer. Methods Engrg.* **46**, 231.
[7] Vladimirov, V.C. (1971). *Equations of mathematical physics*. Nauka, Moscow.
[8] Lai, Y.M., Wu, Z.W., Zhu, Y.L. and Zhu, L.N. (1998). Nonlinear analysis for the coupled problem of temperature, seepage and stress fields in cold-region tunnels. *Tunneling and Underground Space Technology* **13**(4), 435.
[9] COMITE EURO-INTERNATIONAL DU BETON (CEB), "Fire design of concrete structure," CEB, *Lausanne, July* (1991).
[10] Xiao J. and König G. "Study on concrete at high temperature in China-an overview". *Fire Safety Journal*, **39**, 89–103(2004)10.
[11] Lu ZhD., "A Research on fire response of reinforced concrete beams" *PhD thesis, Tongji University*, (1989)
[12] Ding W.," Research and evaluation on repair of RC frame after fire by expert system", *Master thesis, Tongji University*, (1991).
[13] Wu B., Yuan J. and Wang GY., "Experimental study on the mechanical properties of HSC after high temperature" *Chin J Civil Eng.*,**33**, 8–15 (2000).
[14] Hu B. L., Song Y. P. and Zhao GF., "Test on strength and deformation of concrete under complex stress at elevated temperature", *Build Sci Res Sichuan,* **20**, 47–50 (1994).
[15] Li W. and Guo ZhH., "Experimental investigation on strength and deformation of concrete under high temperature". *Chin. J Build Struct.* **14**, 8–16.

Natural convection in porous media under cross temperature and concentration gradients with Boundary Element Method

J. Kramer[1], R. Jecl[1] & L. Škerget[2]
[1] Faculty of Civil Engineering, University of Maribor, Slovenia
[2] Faculty of Mechanical Engineering, University of Maribor, Slovenia

Abstract

Double-diffusive natural convection in a horizontal enclosure is studied numerically with the use of Boundary Element Method. The enclosure is heated and cooled along vertical walls, while horizontal walls are maintained at different concentration values. The Brinkman model is used to simulate the fluid flow through porous media, the density variation is taken into account by the Boussinesq approximation. The governing parameters of the problem are the Darcy number, porous Rayleigh number, Lewis number and buoyancy coefficient. Different flow regimes for thermal and solutal dominated flows are demonstrated for different values of governing parameters. The results are compared to those in published studies, obtained with other approximative methods.
Keywords: Boundary Element Method, porous medium, double-diffusive natural convection, Brinkman model.

1 Introduction

Double-diffusive natural convection in a saturated porous media is of fundamental importance in many engineering branches e.g. civil engineering, mechanical engineering, chemical engineering. Several problems in nature or industry can be modeled as convective flow in enclosure filled with porous media such as underground spreading of pollutants, geophysical systems, fuel storage installations. In general two main configurations have been considered in the literature: heat and mass gradients are imposed horizontally along the enclosure or heat and mass gradients are imposed vertically along the height of the enclosure. Only few recently

WIT Transactions on Modelling and Simulation, Vol 47, © 2008 WIT Press
www.witpress.com, ISSN 1743-355X (on-line)
doi:10.2495/BE080071

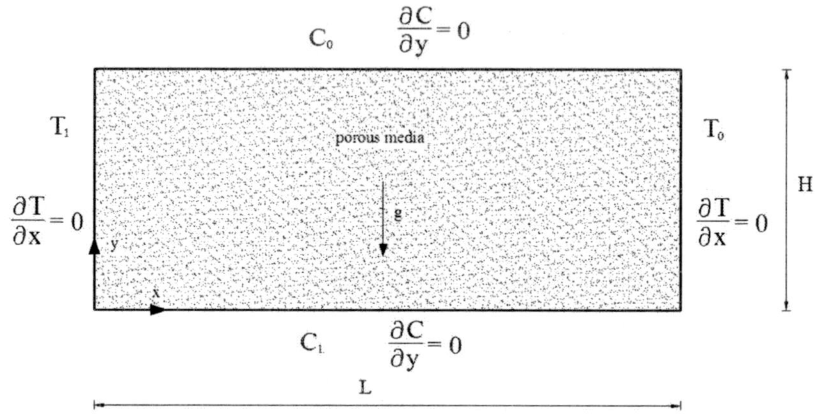

Figure 1: Geometry with boundary conditions for convection with horizontal temperature and vertical solutal gradient.

published studies are dealing cross relation between the temperature and solutal gradient (temperature or concentration gradient imposed horizontally and concentration or temperature gradient imposed vertically) [1–3]. In the studies is shown, that the competition between the thermal and solutal force produce very complex flow patterns.

In the present work the numerical results for heat and solute transfer in a horizontal cavity of aspect ratio $A = 2$, subjected to cross temperature and solute gradients are presented. In the analysis the Brinkman model is adopted while the governing set of equations is solved with use of the extended Boundary Element Method. The results are obtained for different governing parameters, the accuracy of proposed numerical scheme is stated with comparison to some published results.

2 Mathematical model

The physical model with corresponding boundary conditions is shown in Figure 1. The problem under consideration is a two dimensional enclosure with aspect ratio $A = L/H = 2$ filled with fully saturated porous media. The vertical walls are subjected to temperature differences (T_1 and T_0 respectively) while the horizontal walls are subjected to fixed concentrations (high concentration at the bottom C_1 and low concentration at the top C_0). The main assumptions are that the flow is incompressible, steady, laminar, the fluid is Newtonian. The solid matrix is nondeformable, homogenous, isotropic and in thermal equilibrium with the fluid phase.

Transport phenomena in porous media is mathematically described with conservation equations for mass, momentum, energy and species, which are written at the macroscopic level considering the fact that only a part of the volume, expressed with porosity ϕ, is available for the fluid flow [4]. The general set of equations is given as:

- Continuity:

$$\frac{\partial v_i}{\partial x_i} = 0, \tag{1}$$

- Momentum:

$$\frac{1}{\phi}\frac{\partial v_i}{\partial t} + \frac{1}{\phi^2}\frac{\partial v_j v_i}{\partial x_j} = -\frac{1}{\rho_0}\frac{\partial p}{\partial x_i} + Fg_i - \frac{\nu}{K}v_i + \frac{\partial}{\partial x_j}\left(2\frac{\nu}{\phi}\dot{\varepsilon}_{ij}\right), \tag{2}$$

- Energy:

$$\frac{\partial}{\partial t}[\phi c_f + (1-\phi)c_s]T + c_f\frac{\partial v_j T}{\partial x_j} = \frac{\partial}{\partial x_j}\left(\lambda_e\frac{\partial T}{\partial x_j}\right), \tag{3}$$

- Species:

$$\phi\frac{\partial C}{\partial t} + \frac{\partial v_j C}{\partial x_j} = \frac{\partial}{\partial x_j}\left(D\frac{\partial C}{\partial x_j}\right). \tag{4}$$

The parameters, used above are: v_i volume-averaged velocity, x_i the i-th coordinate, ϕ porosity, t time, ρ density, ν kinematic viscosity, $\partial p/\partial x_i$ the pressure gradient, g_i gravity and K permeability of porous media. Parameters $c_f = (\rho c)_f$ and $c_s = (\rho c)_s$ are the heat capacities for the fluid and solid phases, respectively, T is temperature, λ_e the effective thermal conductivity of the porous media given as $\lambda_e = \phi\lambda_f + (1-\phi)\lambda_s$, where λ_f and λ_s are thermal conductivities for the fluid and solid phases, respectively. In the final equation C stands for concentration, and D for mass diffusivity. In the momentum equation (2), the additional Brinkman viscous term is included, which express the viscous resistance or viscous drag force exerted by the solid phase on the flowing fluid at their contact surfaces. The density of the fluid depends only on temperature and concentration variations and is described with Boussinesq approximation as:

$$\rho(T, C) = \rho_0[1 - \beta_T(T - T_0) - \beta_C(C - C_0)], \tag{5}$$

where β_T and β_C are volumetric thermal and concentration expansion coefficients respectively. The equations are coupled through the term F in the momentum equation, which is the density difference function and includes the expression (5) written above.

The boundary conditions for the governing equations can be written as:

- left wall

$$x = 0; \quad v_x = v_y = 0, \quad T = 1, \quad \frac{\partial C}{\partial x} = 0, \tag{6}$$

- right wall

$$x = 2; \quad v_x = v_y = 0, \quad T = 0, \quad \frac{\partial C}{\partial x} = 0, \tag{7}$$

- bottom wall

$$y = 0; \quad v_x = v_y = 0, \quad \frac{\partial T}{\partial y} = 0, \quad C = 1, \tag{8}$$

- top wall

$$y = 2; \quad v_x = v_y = 0, \quad \frac{\partial T}{\partial y} = 0, \quad C = 0. \tag{9}$$

The non-slip boundary conditions are prescribed at the impermeable walls of the enclosure. Along the vertical walls the constant temperatures and adiabatic conditions are assumed. The constant species concentrations are assumed along the horizontal walls and zero mass fluxes at the vertical walls of the enclosure.

3 Numerical procedure

The governing set of equations (1), (2), (3) and (4) is solved using the extended Boundary Element Method, called also the Boundary Domain Integral Method (BDIM). Since the integral equations are given in terms of variables on the integration boundaries, as well as within the integration domain, the discretization of the solution boundary and domain is required. The whole numerical scheme was already presented at BEM/MRM the details are given in [5], [4]. The main steps to obtain a solution of the described problem with BDIM are:

- introduction of modified velocity in all governing equations: $v_i' = v_i/\phi$,
- introduction of heat diffusivity $a_T = \lambda_e/\rho c_f$ and heat capacity $\sigma = \phi + (1-\phi)(\rho c)_s/(\rho c)_f$ in the energy equation,
- separation of velocity field into a constant and variable computational parts $v_i' = \bar{v'}_i + \tilde{v'}_i$,
- separation of material properties (kinematic viscosity, heat diffusivity, species diffusivity) into a constant and variable parts as: $\nu = \bar{\nu} + \tilde{\nu}$, $a_T = \bar{a}_T + \tilde{a}_T$, $D = \bar{D} + \tilde{D}$,
- transformation of governing equations with use of velocity-vorticity formulation (with introduction of vorticity vector as a curl of velocity field $\omega' = e_{ij}\partial v_j'/\partial x_i$ for the case of planar geometry), which consequently separates the computational scheme into a kinematic and kinetic computational parts,
- derivation of integral equations with use of weighting residual technique or Green fundamental functions, where for the kinematics the elliptic Laplace fundamental solution and for the kinetics the elliptic diffusion-convective fundamental solution are used respectively,
- discretization of solution boundary and domain; all boundary domain integral equations are written for all boundary and domain nodes, where the variation of field functions within each boundary element and internal cell is approximated by the use of appropriate interpolation polynomials,
- solution of coupled matrix system with the use of subdomain technique, where each subdomain consists of four discontinuous 3-node quadratic boundary elements and one 9-node corner continuous quadratic internal cell.

The final implicit matrix system for the solution of velocity field and values of vorticity, temperature and concentration is given for the case of planar geometry as:

- Kinematics:

$$[H]\{v_i'\} = e_{ij}[H_t]\{v_j'\} - e_{ij}[D_j]\{\omega'\}, \tag{10}$$

- Vorticity kinetics:

$$\begin{aligned}[H]\{\omega'\} &= \frac{1}{\bar{\nu}}[G]\left\{\nu\frac{\partial\omega'}{\partial n} - \omega' v_n' + e_{ij}g_j F n_j + f_j n_j\right\} \\ &+ \frac{1}{\bar{\nu}}[D_j]\left\{\omega'\tilde{v}_j' - e_{ij}g_j F - \tilde{\nu}\frac{\partial\omega'}{\partial x_j} - f_j\right\} \\ &+ \frac{1}{\bar{\nu}\Delta t}[B]\left\{\frac{\nu\phi}{K}\omega' + \omega'_{F-1}\right\},\end{aligned} \tag{11}$$

- Energy kinetics

$$\begin{aligned}[H]\{T\} &= \frac{\phi}{\sigma\bar{a}_p}[G]\left\{\frac{\sigma a_p}{\phi}\frac{\partial T}{\partial n} - T v_n'\right\} - \\ &- \frac{\phi}{\sigma\bar{a}_p}[D_j]\left\{\frac{\sigma\tilde{a}_p}{\phi}\frac{\partial T}{\partial x_j} - T\tilde{v}_j'\right\} + \frac{1}{\bar{a}_p\Delta t}[B]\{T\}_{F-1},\end{aligned} \tag{12}$$

- Species kinetics

$$\begin{aligned}[H]\{C\} &= \frac{\phi}{\bar{D}}[G]\left\{\frac{D}{\phi}\frac{\partial C}{\partial n} - C v_n'\right\} \\ &- \frac{\phi}{\bar{D}}[D_j]\left\{\frac{\tilde{D}}{\phi}\frac{\partial C}{\partial x_j} - C\tilde{v}_j'\right\} + \frac{\phi}{\bar{D}\Delta t}[B]\{C\}_{F-1}.\end{aligned} \tag{13}$$

The term f_j in the vorticity equation describes any contribution arising on account of nonlinear material properties and is given as $f_j = \tilde{\nu}(\vec{\nabla} \times \dot{\varepsilon}_{ij})$. The matrix $[H]$, $[H_t]$, $[G]$, $[D_j]$ and $[B]$ are the influence matrices composed of those integrals which took over the individual boundary elements and internal cells.

4 Results and discussion

The presented numerical scheme was used to solve a problem of double-diffusive natural convection due to cross temperature and concentration gradients as shown in the chapter 2. The governing parameters that define the fluid flow, heat and solute transfers are:

- aspect ratio $A = L/H$, where L and H are length and height of the enclosure respectively,
- Darcy number $Da = K/H^2$,
- porous Rayleigh number $Ra = Kg_i\beta_T, \Delta TH/a_T\nu$, where ΔT is the temperature difference, a_T thermal diffusivity and ν kinematic viscosity,

- Lewis number $Le = a_T/D$,
- buoyancy coefficient $N = \beta_C \Delta C / \beta_T \Delta T$, where ΔC is concentration difference, all other parameters are already described.

In the presented example the results for enclosure of aspect ratio $A = 2$, $Ra = 100$, $Da = 10^{-3}$, $Le = 10$ and different values of buoyancy coefficient are presented. A non-uniform computational mesh 20×10 was used with a ratio between the longest and shortest elements of $r = 6$. Time-steps of $\Delta t = 10^{-}1$ were employed for all cases, and the convergence criterion is determined as $\varepsilon = 5 \times 10^{-6}$. In the Figure (2) left the streamlines, isotherms and concentration lines for $N = 0$ and right for $N = -1.5$ are shown and in the Figure (3) for $N = -2$ left and $N = -4$ right. For the case $N = 0$, where flow is driven by thermal buoyancy force, the flow structure consists of one main circulation occupying the entire enclosure. At the core of the enclosure the concentration gradient reversal is evident, due to strong flow recirculation. From the temperature filed is evident that the rate of heat transfer is high at the bottom of the hot wall and upper part of the cold wall. The rate of heat transfer decreases along the hot wall. With increase of buoyancy coefficient N in negative sense the strength of the flow circulation decreases and starts to bifurcate into two weak circulations. Physical explanation for the cases when $|N| > 0$ is that heavy fluid particles (high concentration) that advected to the upper part of the enclosure reach the other end and form recirculation. As N decreases the

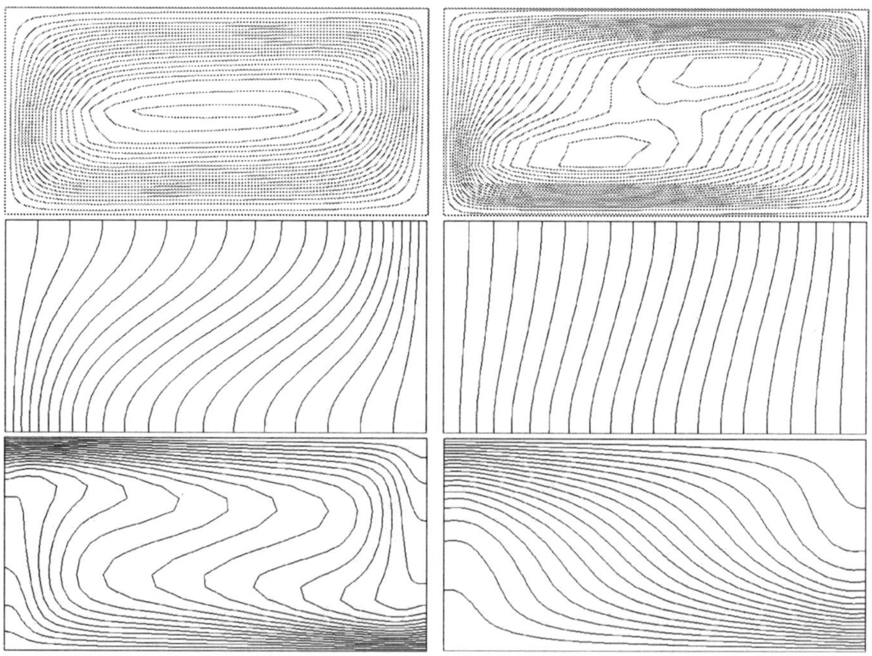

Figure 2: Streamlines, isotherms and isoconcentrations for $A = 2$, $Ra = 100$, $Da = 10^{-3}$, $Le = 10$ and $N = 0$ (left) and $N = -1.5$ (right).

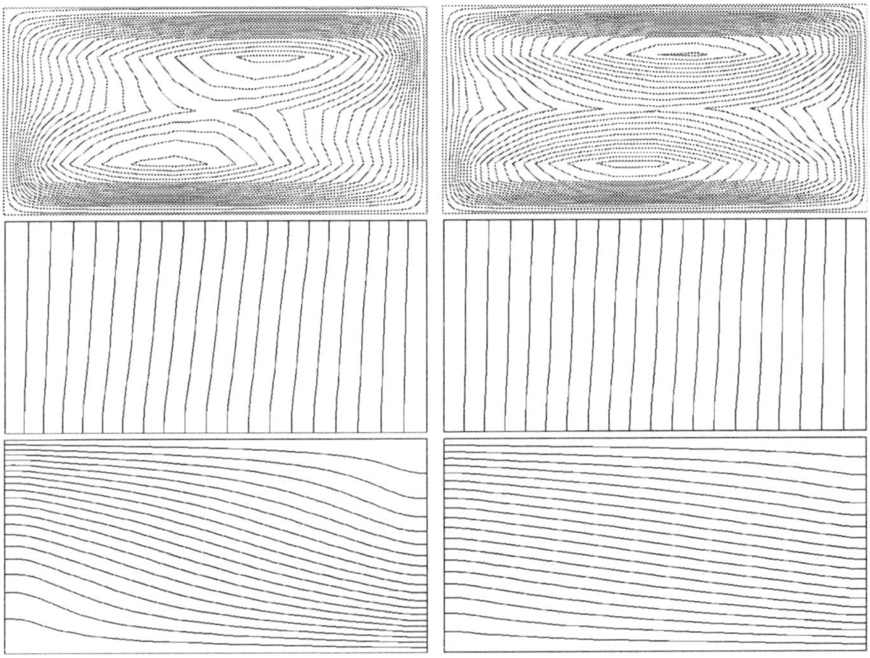

Figure 3: Streamlines, isotherms and isoconcentrations for $A = 2$, $Ra = 100$, $Da = 10^{-3}$, $Le = 10$ and $N = -2$ (left) in $N = -4$ (right).

heavy particles sink before they arrive at the other end of the enclosure forming two circulations. With increase of N also the concentration reversal and isotherms distortion diminish. For $N = -4$ heat transfer takes place mainly by conduction, which is evident from the isotherm distribution. The sharp corners in the streamline presentations in Figure (3) are due to small grid density (20×10 subdomains). For better graphical presentations denser grids have to be used. Table (1) shows results for the values of Nusselt and Sherwood numbers for the examples shown in Figures 2 and 3. As already stated from the graphical presentations, the heat is for small values of N transfered by conduction, while the solute is transported by the convection mechanism. The results are compared to the reference [3], where the same problem is solved for planar and spatial geometry with use of finite difference method (FDM). The agreement between the results is good, which proves the accuracy of mathematical model and BDIM.

5 Conclusion

In the paper the results for double-diffusive natural convection in a rectangular enclosure subjected to cross gradients of temperature and concentration, obtained with use of extended Boundary Element Method are presented. The results show that the competition between solutal and thermal force produce complex flow pat-

Table 1: Nu and Sh values for different N and $A = 2$, $Ra = 100$, $Le = 10$, $Da = 10^{-3}$.

N		−4	−2	−1.5	−1
Nu	Present results	**0.49**	**0.50**	**0.50**	**0.52**
	Ref. [3]	–	0.5	0.5	0.55
Sh	Present results	**1.04**	**1.17**	**1.30**	**1.71**
	Ref. [3]	–	1.2	1.3	1.9

terns. The dependency of buoyancy coefficient on the convective flow in the enclosure is investigated and it is stated that with increase of buoyancy coefficient N in negative sense the convective motion slows down and approaches the pure diffusion limit. The results are compared to a published study, where the FDM was used to solve the same problem. Very good agreement between the results is stated, which indicates that the BDIM could be efficiently used for solving the double-diffusive natural convection problem in porous media.

References

[1] Mohamad, A.A. & Bennacer, R., Natural convection in a confined saturated porous medium with horizontal tempertaure and vertical solutal gradients. *Int J Therm Sci*, **40**, pp. 82–93, 2001.

[2] Kalla, L., Vasseur, P., Beji, H. & Duval, R., Double diffusive convection within a horizontal porous layer salted from the bottom and heated horizontally. *Int Comm Heat Mass Transfer*, **28**, pp. 1–10, 2001.

[3] Mohamad, A.A. & Bennacer, R., Double diffusion, natural convection in an enclosure filled with saturated porous medium subjected to cross gradients; stably stratified fluid. *Int J Heat Mass Transfer*, **45**, pp. 3725–3740, 2002.

[4] Kramer, J., Jecl, R. & Škerget, L., Boundary element method for double diffusive natural convection in horizontal porous layer. *Boundary elements and other mesh reduction methods XXIX*, 2007.

[5] Škerget, L., Hriberšek, M. & Kuhn, G., Computational fluid dynamics by boundary-domain integral method. *Int J Numer Meth Engng*, **46**, p. 1291.1311, 1999.

Section 3
Electrical engineering and electromagnetics

Analysis of log-periodic dipole arrays with boundary elements

D. Poljak[1], V. Doric[1], M. Birkic[2] & D. Kosor[3]
[1]*Department of Electronics, University of Split, Croatia*
[2]*Technical Department, Air Traffic Center Pula, Croatia*
[3]*Technical Department, Air Traffic Center Split, Croatia*

Abstract

The paper deals with a frequency domain analysis of a log-periodic dipole array (LPDA) antenna using boundary element analysis. The mathematical model is based on the set of corresponding coupled Pocklington integro-differential equations. Once the current distribution along the wires is obtaining, it is possible to determine the radiation characteristics of LPDA. The set of coupled Pocklington integro-differential equations is handled via the Galerkin-Bubnov scheme of the Indirect Boundary Element Method (GB-IBEM). The field irradiated by LPDA is also assessed using BEM formalism.
Keywords: log-periodic dipole arrays, set of Pocklington equations, boundary elements, frequency domain analysis.

1 Introduction

A log-pariodic antenna is a radiating system having a structure providing that its impedance and radiation properties repeat periodically as the logarithm of frequency [1, 2]. As the variations over a wide frequency range of interest are negligible, log-periodic antennas are usually considered to be frequency-independent.These antennas operate in the VHF and UHF bands thus covering the frequency range from 30MHz to 3GHz.

The log-periodic dipole array (LPDA) antenna is a series-fed array of parallel straight wires having successively increasing lengths outward from the feeding point at the apex. There are smaller elements in front of and larger elements behind each dipole thus producing a directional primary pattern in the same way as a Yagi antenna. Contrary to the Yagi-Uda array where only one element of the array is directly fed, while the other wires operate in a parasitic mode, all the elements of the LPDA are connected to the feeder.

WIT Transactions on Modelling and Simulation, Vol 47, © 2008 WIT Press
www.witpress.com, ISSN 1743-355X (on-line)
doi:10.2495/BE080081

Log-periodic dipole array antennas are driven by attaching dipole arms to two parallel conductors which represent a transmission line [1]. The transmission line feeder is crossed between each dipole thus reversing the direction of fire. The LPDA antennas are found to be easy to optimize, while the crossing of the feeder between each dipole element leads to a mutual cancellation of backlobe components from the individual elements yielding to a very low level of backlobe radiation (around 25dB below main lobe gain at HF and 35dB at VHF and UHF).

To maintain the geometrical pattern of the LPDA to achieve a trully log-periodic configuration, an infinite configuration would be required. However, the practical broadband radiator configuration is truncated at both ends which limits the frequency of operation to a given bandwidth. The cutoff frequencies of the truncated structure is determined by the electrical lengths of the largest and shortest elements of the structure [1, 2]. It is worth mentioning that the use of logarithmic antenna arrays is very often related with electronic beam steering. An important application of LPDA antennas is in air traffic. Landing is one of teh most important issues of every successful flight. Although the majority of landings can be conducted solely with visual cues, aircraft must frequently land in weather conditions that require electronic assistance to the pilot or to the autopilot.

Log-periodic dipole antenna is an essential part of localizer antenna arrays. A typical localizer antenna system is a part of the electronic systems known as Instrumental Landing System (ILS). Localizer shapes a radiation pattern in a way to provide lateral guidance to the aircraft beginning its descent, intercepting the projected runway center line, and then making a final approach.

This work deals with a boundary element modeling of LPDA antennas. The formulation in the frequency domain is based on a set of coupled Pocklington integro-differential equations for radiation of multiplethin wires above a lossy ground. The set of Pocklington equations is solved via the Galerkin-Bubnov variant of the indirect Boundary Element Method (GB-IBEM) [4]. Once the induced currents along the LPDA are determined the radiation pattern is assessed using BEM formalism, as well.

2 Formulation

Modeling of multiple wire configurations above a lossy ground is an important part in antenna design and electromagnetic compatibility (EMC) studies [2, 5–8]. This section firstly deals with an assessment of the current induced along multiple wires above a lossy ground at different heights. Once the currents along the wire array have been obtained, the equations for related radiated field are presented [8]. In this case, the geometry of interest is LPDA shown in Fig 2. The length of any other wire is obtained by multiplying the previous length and factor τ [2]:

$$\tau = \frac{L_{n+1}}{L_n} \tag{1}$$

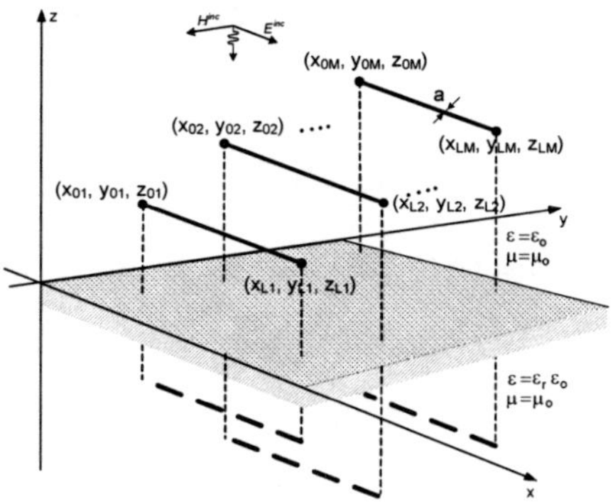

Figure 1: Wires above a lossy ground at different heights.

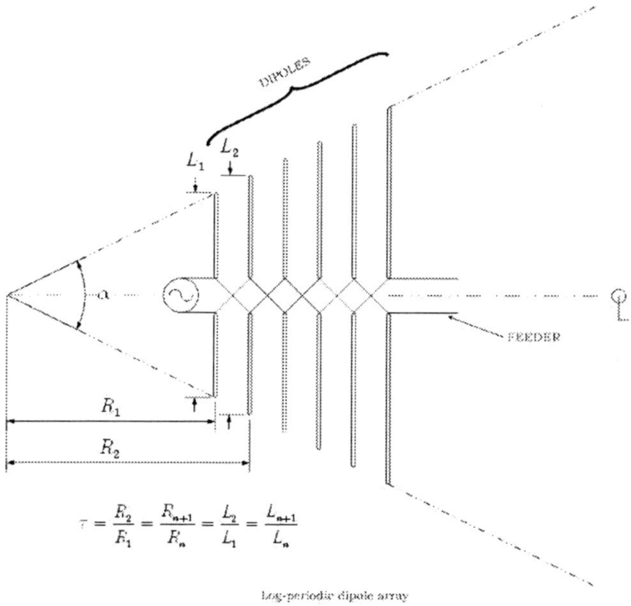

Figure 2: LPDA geometry.

The current distribution along the multiple wire structure shown in Fig 1 is governed by the set of Pocklington integro-differential equations for half-space problems given by [8]:

$$E_x^{exc} = -\frac{1}{j4\pi\omega\varepsilon_0}\sum_{n=1}^{M}\int_{-L_n/2}^{L_n/2}\left[\frac{\partial^2}{\partial x^2}+k_1^2\right]\left[g_{0mn}(x,x') - R'_{TM}\, g_{imn}(x,x')\right] I_n(x')dx' \tag{2}$$

$$m = 1, 2, \ldots M$$

where M is the total number of wires and $I_n(x')$ is the unknown current distribution induced on the n-th wire. Furthermore, $g_{0mn}(x,x')$ denotes the free space Green function:

$$g_{0mn}(x,x') = \frac{e^{-jk_1R_{1mn}}}{R_{1mn}} \tag{3}$$

while $g_{imn}(x,x')$ is, in accordance to the image theory, given by:

$$g_{imn}(x,x') = \frac{e^{-jk_1R_{2mn}}}{R_{2mn}} \tag{4}$$

where k_1 is the propagation constant of free space:

$$k_1^2 = \omega^2 \mu_0 \varepsilon_0 \tag{5}$$

and R_{1mn} and R_{2mn} are distances from the source point and from the corresponding image, respectively to the observation point of interest.

The influence of an imperfectly conducting lower medium is taken into account via the Fresnel plane wave reflection coefficient [RC] [5–8]:

$$R'_{TM} = \frac{\underline{n}\cos\theta' - \sqrt{\underline{n} - \sin^2\theta'}}{\underline{n}\cos\theta' + \sqrt{\underline{n} - \sin^2\theta'}} \tag{6}$$

where θ' and $\underline{n}$ are given by:

$$\theta' = arctg\frac{|x - x'|}{2h} \quad m = n \quad \theta' = arctg\frac{R_{1mn}}{2h} \quad m \neq n, \; \underline{n} = \frac{\varepsilon_{eff}}{\varepsilon_0} \tag{7}$$

and ε_{eff} is the complex permittivity of the ground:

$$\varepsilon_{eff} = \varepsilon_r \varepsilon_0 - j\frac{\sigma}{\omega} \tag{8}$$

The electric field (normal incidence) at height z above finitely conducting half-space can be written, as follows:

$$E^{inc} = E_0 e^{-jk_1(z_U - z)} \tag{9}$$

where E_0 denotes the field amplitude, and z_U is the position of the highest antenna among M wires in total: $z_U = \max(z_1, z_2, \ldots z, \ldots z_M)$.

The earth- reflected field component is given by:

$$E^{ref} = E_0 R_{TM} e^{-jk_1(z_U + z)} \tag{10}$$

The total field above ground is composed from both incident and reflected field component:

$$E^{exc} = E^{inc} + E^{ref} \tag{11}$$

The electric field components irradiated by the array are given, as follows [8]:

$$E_x = \frac{1}{j4\pi\omega\varepsilon_0}\sum_{n=1}^{M}\left[-\int_{-L_n}^{L_n}\frac{\partial I_n(x')}{\partial x'}\frac{\partial G_{nm}(x,x')}{\partial x'}dx' + k^2\int_{-L_n}^{L_n}G_{nm}(x,x')I_n(x')dx'\right] \tag{12}$$

$$E_y = \frac{1}{j4\pi\omega\varepsilon_0}\sum_{n=1}^{M}\int_{-L_n}^{L_n}\frac{\partial I_n(x')}{\partial x'}\frac{\partial G_{nm}(x,x')}{\partial y}dx' \tag{13}$$

$$E_z = \frac{1}{j4\pi\omega\varepsilon_0}\sum_{n=1}^{M}\int_{-L_n}^{L_n}\frac{\partial I_n(x')}{\partial x'}\frac{\partial G_{nm}(x,x')}{\partial z}dx' \tag{14}$$

where the corresponding Green function is given by:

$$G_{nm}(x,x') = g_{0nm}(x,x') - R_{TM}\, g_{inm}(x,x') \tag{15}$$

3 Boundary element procedures

The Galerkin boundary element procedure for the treatment of the system of integro-differential equations (1), using linear approximation of the solution, starts by applying the standard representation of the unknown current along a segment [4, 8]:

$$I(x') = I_{1i}\frac{x_{2i}-x'}{\Delta x} + I_{2i}\frac{x'-x_{1i}}{\Delta x} \tag{16}$$

where I_{1i} and I_{2i} are current values at the local nodes of the i-th boundary element, with coordinates x_{1i} and x_{2i}, $\Delta x = x_{2i} - x_{1i}$ denotes a dimension of the element.

Performing certain mathematical manipulations and the boundary element discretisation of the wires results in the following matrix equation [4, 8]:

$$\sum_{k=1}^{N_e}[Z]_{pk}\{I\}_k = \{V\}_p \quad p=1,2,\ldots,M \tag{17}$$

where N_e is the total number of elements along the actual multiple wire configuration, and $[Z]_{pk}$ is the interaction matrix representing the mutual impedance between each segment on the i-th (source) wire to every segment on the j-th (observation) wire [4, 8]:

$$[Z]^e_{pk} = -\int_{\Delta l_p}\int_{\Delta l_k}\{D\}_p\{D'\}^T_k\, g_{ji}(x,x')dx'dx + k^2\int_{\Delta l_p}\int_{\Delta l_k}\{f\}_l\{f'\}^T_k\, g_{ji}(x,x')dx'dx \tag{18}$$

Vectors $\{f\}$and $\{f'\}$ contain shape functions $f_n(x)$ and $f_n(x')$, while $\{D\}$ and $\{D'\}$ contain their derivatives. The right side vector $\{V\}_p$ represents the voltage along the p-th segment and it is given as follows [4, 8]:

$$\{V\}_p = -j4\pi\omega\varepsilon_0\int_{\Delta l_p}E_x^{inc}(x)\{f\}_p\,dx \tag{19}$$

Transmitting mode implies that the vector (19) differs from zero only in the feed-gap area of each wire, Fig 2, and the excitation field can be written, as follows:

$$E_x^{exc} = \frac{V_g}{\Delta l_g} \tag{20}$$

where V_g is the feed voltage and Δl_g is the feed-gap width.

Once computing the currents along the wires the radiated field can be computed. Applying the BEM formalism to field expressions (12)–(14) it follows:

$$E_x = \frac{1}{j4\pi\omega\varepsilon_0} \sum_{n=1}^{M} \sum_{i=1}^{N_j} \left[-\frac{I_{i+1,n} - I_{i,n}}{\Delta x} \int_{x_{i,n}}^{x_{i+1,n}} \frac{\partial G_{nm}(x,x')}{\partial x'} dx' + k^2 \int_{x_{i,n}}^{x_{i+1,n}} G_{nm}(x,x') I_{in}(x') dx' \right]; \quad (21)$$

$m = 1, 2, ..., M$

$$E_y = \frac{1}{j4\pi\omega\varepsilon_0} \sum_{n=1}^{M} \sum_{i=1}^{N_j} \frac{I_{i+1,n} - I_{i,n}}{\Delta x} \int_{x_{i,n}}^{x_{i+1,n}} \frac{\partial G_{nm}(x,x')}{\partial y} dx'; \quad m = 1, 2, ..., M \quad (22)$$

$$E_z = \frac{1}{j4\pi\omega\varepsilon_0} \sum_{n=1}^{M} \sum_{i=1}^{N_j} \frac{I_{i+1,n} - I_{i,n}}{\Delta x} \int_{x_{i,n}}^{x_{i+1,n}} \frac{\partial G_{nm}(x,x')}{\partial z} dx'; \quad m = 1, 2, ..., M \quad (23)$$

where N_j denotes the total number of boundary elements on the j-th wire.

The integrals in (21)–(23) are evaluated using Gaussian quadrature. To avoid the problem of the Green function quasi-singularity its first derivative is approximated by means of a central finite difference approximation [4, 8].

4 Computational example

The computational example is related to the LPDA composed from 12 dipoles insulated in free space. The radius of all wires is a=0.004m while the length of

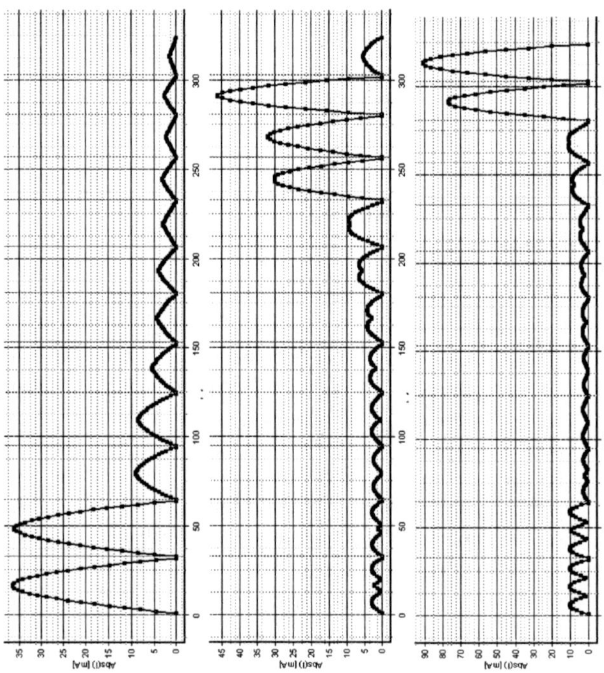

Figure 3: Absolute value of current distribution along 12 dipoles versus BEM nodes at f=100MHz, f=250MHz and f=300MHz.

wires are determined by the length of 1st wire L_1=1.5m, and factor τ=0.9. All dipoles are fed by the voltage generator V_g=1V with variable phase (each time phase is changed for 180°). The operating frequency is varied from 100 MHz to 300 MHz. Figure 3 shows the absolute value of currents induced along all 12 dipoles versus BEM nodes at *f*=100MHz, *f*=250MHz and *f*=300MHz. Figs 4 to 6 show the related field patterns at *f*=100MHz, *f*=250MHz and *f*=300MHz.

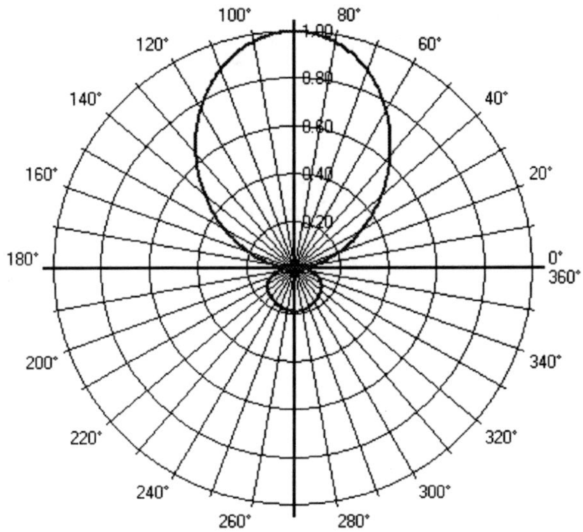

(XY plane)

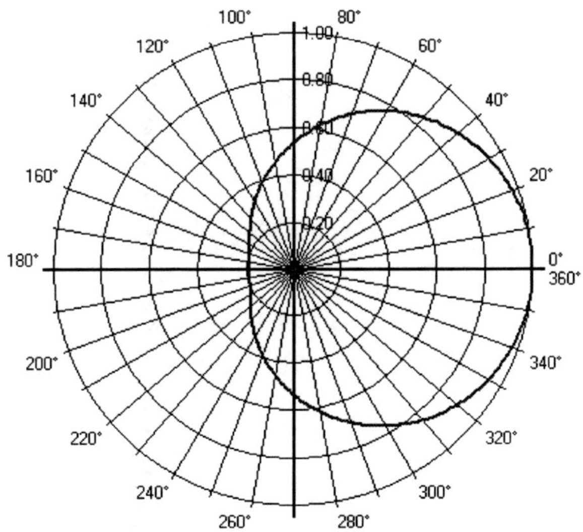

(YZ plane)

Figure 4: Radiation pattern at *f*=100MHz.

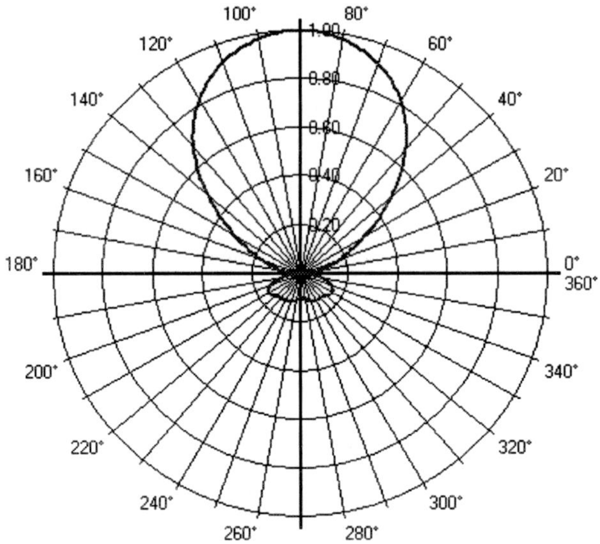

(XY plane)

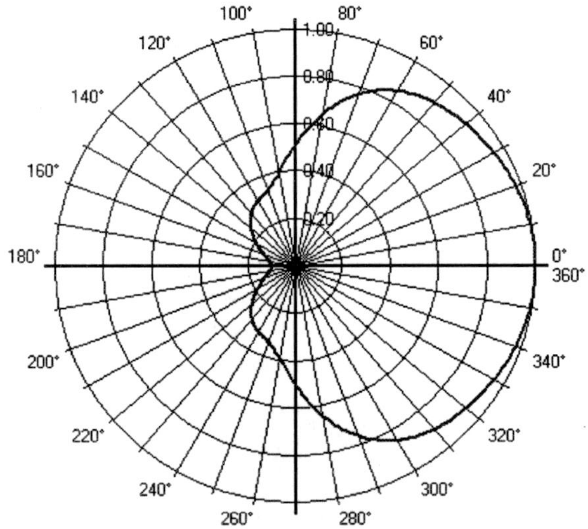

(YZ plane)

Figure 5: Radiation pattern at f=250MHz.

5 Concluding remarks

The frequency domain analysis of the log-periodic dipole array (LPDA) with boundary elements is undertaken in this work. The formulation is based on the set of coupled integro-differential equations of the Pocklington type. The induced currents along the elements of the array are obtained as the BEM solution of the Pocklington equations. Once obtaining the current distribution along the elements of LPDA the related far field patterns are computed.

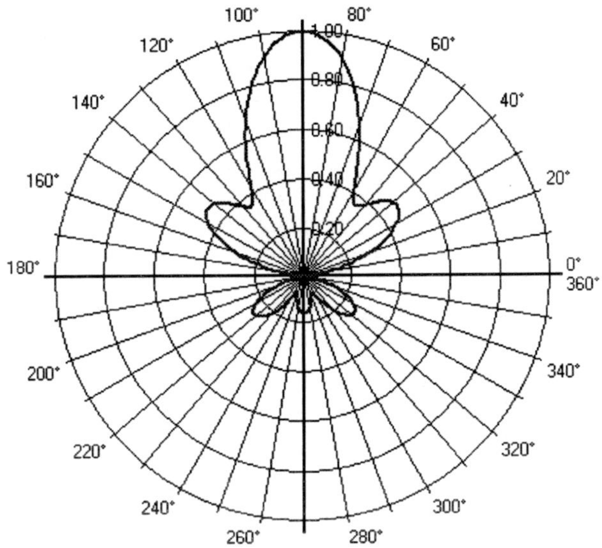

(XY plane)

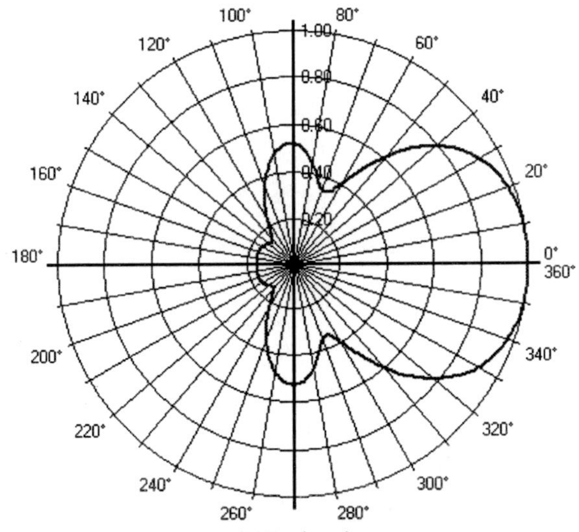

(YZ plane)

Figure 6: Radiation pattern at f=300MHz. (YZ plane).

References

[1] R.A. Burberry, VHF and UHF Antennas, IEE Electromagnetics Series 35, Peter Peregrinus Ltd, London, UK, 1992.

[2] C.A. Balanis: *Antenna Theory*, 2nd Edition, John Wiley & Sons, Inc., New York, 1997.

[3] Poljak, D., Doric, V., Roje, V., Galerkin-Bubnov Boundary element analysis of the Yagi-Uda array, *24th International Conference on the Boundary Element Method*, BEM XXIV, pp 457-463, Sintra, Portugal, June 2002.

[4] Poljak D., C.A. Brebbia, *Boundary Element Methods for Electrical Engineers*, WIT Press, Southampton-Boston, 2005.

[5] Miller, E.K., Poggio, A.J., Burke, G.J., Selden, E.S., Analysis of wire antennas in the presence of a conducting half-space. Part II. The horizontal antenna in free space, *Canadian Journal of Physics*, **50**, 1972, pp 2614–2627.

[6] Sarkar, T.K., Analysis of arbitrarily oriented thin wire antennas over a plane imperfect ground, *Archiv fur elektronik und ubertragungstechnik*, **31**, 1977, pp 449–457.

[7] Parhami, P. Mittra, R., Wire antennas over a lossy half-space, *IEEE Trans. AP*, **28**, 1980, pp 397–403.

[8] D.Poljak, V.Doric, S.Antonijevic "Computer aided Design of Wire Structures: Frequency and Time Domain Analysis", WIT Press, Southampton-Boston, 2007.

Electromagnetic modelling of foetus and pregnant woman exposed to extremely low frequency electromagnetic fields

C. Gonzalez[1], A. Peratta[1] & D. Poljak[2]
[1] *Wessex Institute of Technology, UK*
[2] *University of Split, Croatia*

Abstract

The paper presents a three dimensional BEM model of a pregnant woman and foetus exposed to extremely low frequency electromagnetic fields. In particular, the case scenario solved is focused on the case of exposure to high voltage power transmission lines. The paper presents the corresponding theory, BEM computational implementation, and results. The latter are expressed in terms of current density, potential and electric field in the different tissues. This work is part of an ongoing research whose overall objectives are to provide accurate estimations of fields and induced currents for assessing exposure of human bodies to electro-magnetic fields.
Keywords: pregnant woman; boundary element method; electric field.

1 Introduction

Exposure levels in the foetus of a pregnant woman are difficult to estimate mainly because of the following three main factors. Firstly, the lack of data on electrical properties at low frequency for the foetus and the surrounding tissues; secondly, the impossibility of collecting *in-vivo* measurements in a real case scenario; and finally, because of the complicated changing geometrical and physical properties of the body along the pregnancy period. Hence, a numerical modelling approach is highly appealing, especially for conducting sensitivity analysis on the electrical properties, which are scarce and scattered in the available literature.

The aim of this paper is to analyse the case of exposure of a pregnant woman to high voltage overhead power-lines by means of a collocation BEM based on an three dimensional anatomical model of the mother and foetus.

WIT Transactions on Modelling and Simulation, Vol 47, © 2008 WIT Press
www.witpress.com, ISSN 1743-355X (on-line)
doi:10.2495/BE080091

The developed code will be used for measuring induced currents and electric fields in the foetus in different scenarios of conductivities at different time stages of pregnancy, and considering different presentations of the foetus inside the maternal matrix. The different stages of pregnancies relevant for the modelling arise not only from the geometrical point of view but also due to the variation of electrical properties of tissues during gestation [1].

In time harmonic EM fields, Maxwell equations can be decoupled when the characteristic size of the model (2 m for the whole human body) is much smaller than the characteristic vacuum wavelength ($\lambda_0 \sim 860$ m at frequency $\nu = 350$ kHz) and the displacement currents can be neglected in comparison to the resistive ones. Therefore, the high voltage LF approach is applicable and the numerical problem summarises into solving the non-homogeneous Laplace equation.

On the other hand, the Boundary Element Method (BEM) [2, 3] is a well established numerical technique which provides accurate solutions in complex geometries. The beauty of BEM is that it takes into account the fundamental solution of the leading partial differential operator, and that the discretisation can be done only in the boundary of the problem. The BEM has been successfully applied for solving problems of biological tissues, other than the human eye, exposed to low and high frequency electromagnetic (EM) fields [4–7].

2 Low frequency EM modelling with boundary elements

When considering biological tissue exposure to high voltage and low currents the most influential field is the electric one. Assuming both conductivity σ and permittivity ε to be constant within a finite region of interest (sub-domain) it is derived from Maxwell's equations that the electric scalar potential φ obeys the non-homogeneous Laplacian-type equation: $\nabla \cdot [(\sigma + i\omega\varepsilon)\nabla\varphi] = 0$, where $e^{i\omega t}$-time dependency convention has been used, being $\omega = 2\pi\nu$ the angular frequency of the incident field, ε the permittivity, and $i^2 := -1$. The corresponding integral equation [2, 3] for the potential $\varphi(\mathbf{x_s})$ in a homogeneous domain is:

$$c_i\varphi(\mathbf{x}_i) + \int_\Gamma \frac{\partial G^*(\mathbf{x}, \mathbf{x}_i)}{\partial \hat{n}} \varphi(\mathbf{x}) d\Gamma - \int_\Gamma G^*(\mathbf{x}, \mathbf{x}_i) \frac{\partial \varphi}{\partial \hat{n}}(\mathbf{x})\, d\Gamma = 0, \qquad (1)$$

where Ω is the integration domain with boundary $\Gamma = \partial(\Omega)$ of outward unit normal $\hat{\mathbf{n}}$, G^* is the Green's function of Laplace equation: $\nabla^2 G^* + \delta(\mathbf{x}_i, \mathbf{x}) = 0$, and $\partial G^*/\partial\hat{n}$ its normal derivative in $\hat{\mathbf{n}}$ direction. Also, proper boundary conditions are applied to $\Gamma = \partial(\Omega)$, i.e. Dirichlet, Neumann or Robin type. In 3D space the Green's function becomes $G^* = 1/(4\pi r)$ where $\mathbf{r} = \mathbf{x} - \mathbf{x}_i$, $r = |\mathbf{r}|$ is the distance between the field ($\mathbf{x} \in \Gamma$) and source ($\mathbf{x}_i \in \Omega$) points, and c_i is a constant dependent on the Cauchy principal value integration of the singularity at the source point. Discretisation with constant elements of eq. (1) yields:

$$c_i\varphi_i + \sum_{j=1}^{N_e} H_{ij}\varphi_j - \sum_{j=1}^{N_e} G_{ij}E_{n,j} = 0, \qquad (2)$$

where φ_j and $E_{n,j}$ denote potential and normal electric field, respectively in j-th element, and $\mathbf{G}$ and $\mathbf{H}$ are the usual BEM single and double layer integrals [3], respectively. The assembly of (1) leads to a linear system of algebraic equations $\mathbf{Ax} = \mathbf{b}$, whose solution provides the unknown φ and E_n values at the boundary. At LF, biological tissues behave as good conductors with conductivity values of the order of 0.5 S/m, and electric permittivity 10^{-10} F/m , i.e. $\varepsilon_r \sim 100$; and the air represents a nearly perfect dielectric. In addition, air has negligible conductivity in comparison with tissues, and the permittivity of most biological tissues is few orders of magnitude greater than ε_0 [8]. Therefore, appropriate matching conditions between air (0) and tissue (1,2) and different tissues, can be written as:

$$\left[\sigma \frac{\partial \varphi_I}{\partial n}\right]_{1,2} = \left[\omega\varepsilon \frac{\partial \varphi_R}{\partial n}\right]_0 \quad ; \left[\sigma \frac{\partial \varphi_I}{\partial n}\right]_1 = \left[\sigma \frac{\partial \varphi_I}{\partial n}\right]_2, \tag{3}$$

respectively; where $\mathbf{j} = -\sigma \nabla \varphi$ is current density and $\mathbf{j} \cdot \hat{\mathbf{n}}$ is preserved.

3 Physical model

The physical model involves two challenging aspects, the changing geometrical and physical data of the relevant tissues for mother and foetus. Different stages of pregnancy are modelled separately in order to consider the changes throughout gestation. These changes are related not only to the volume, mass and geometry of the maternal body and foetus, but also to the electrical properties of the participating tissues.

The background information for foetal and embryo development was obtained from ref. [9]. The embryonic period (3rd to 8th weeks) is the time when all internal and external structures develop in the embryo. During this critical period, the exposure of an embryo to certain agents such as external electromagnetic fields may cause major congenital malformations.

The end of the embryonic stage occurs by the end of eighth week and then the foetal period begins. During the fetal period, the growth, development and maturation of the structures that have been already formed takes place. The definition of the model in this work reflects the four different stages of pregnancy spread along gestation, and the timing correspond to the 8th, 13th, 26th and 38th gestational weeks.

The conductivity data for the foetus is scarce and scattered in the literature. For the maternal abdomen, the division into sub-domains is based on the different properties of the tissues. The amniotic fluid (AF) has the highest conductivity which varies depending on the period of gestation [1,8]. Kidney, muscle bone cortical, bladder, spleen, cartilage and skin have all conductivity values very close to 0.1 S/m, ovary and cartilage have conductivity $\sim$ 0.2 S/m. Therefore, all these tissues can be grouped into one sub-domain, namely maternal tissue. The uterus conductivity is 0.23 S/m, which is very similar to the conductivity of the maternal tissue. According to ref. [1], the placenta is assumed to have the same conductivity as the blood. and considered as part of the maternal-tissue sub-domain.

Consequently, the maternal abdomen is divided into three sub-domains namely "maternal tissue", "amniotic fluid", contained within the uterus, and "foetus".

The geometrical information for the foetus model was built with the help of CT images at different stages [10] and data on the anatomy of the mother and foetus were extracted from [11]. The anthropometric measurements for the size of the foetus are based on the crownrump length (CRL), and the biparietal diameter (BPD) [11], whose growth is almost linear in the early weeks of pregnancy, but there is a progressive reduction in growth rate, especially during the final weeks. Reference values of the surface area of the foetal body [11] were also adopted for defining the model. During the foetal period length and weight do not change in the same way. Foetal length change is greatest in the second trimester, while foetal weight change is greatest in the final weeks of development.

Furthermore, the foetus is free to move inside the maternal abdomen, principally until the 24th week. Since then, the movement is more constrained. In obstetrics, the foetal orientation and position are normally described in terms of the *foetal lie, presentation attitude* and *position*. The *foetal lie* describes the orientation of the longitudinal axe of the foetus in relation to the longitudinal axe of the mother. If the longitudinal axe of the foetus is parallel to the longitudinal axe of the mother, the foetus is in *longitudinal lie*, while if it is perpendicular or oblique, the foetus is in *transverse* or *oblique lie* respectively. Longitudinal lie occurs in the 95% of the cases.

The fetal attitude describe the relative position of different parts of the body of the foetus in relation with his own body. For example, in the most normal fetal attitude, referred as well as *the fetal position*, the head is tucked down to the chest, with arms and legs drawn in towards the centre of the chest.

The *presentation* of the foetus refers to his orientation in relation with the birth channel. The normal presentation is cephalic, with the head oriented to the birth channel. When the foetus is in cephalic presentation and fetal position, then the presentation is referred to as *vertex presentation*. This presentation is the most common at delivery and occurs in the 96% of the births. Another presentation that occurs in the 3.5% of the births is the *breech presentation* when the buttocks are oriented towards the birth channel. The last and less frequent presentation (0.5%) is the *shoulder presentation* associated with transverse lie.

Figure 1 shows a view of the model developed for the 26 weeks foetus in the cephalic and breech presentations and a general 3D view of one of the models (1.7 m tall) at 26 weeks of pregnancy with the foetus in cephalic presentation. The data for the maternal geometry and its variation along pregnancy was obtained from [11]. Roughly, during gestation there is a change in the volume of AF, and a general increase of mass distributed over the maternal body. Figure 1 shows a The scenarios in this work are prepared in order to study different scenarios of electrical conductivity (see Table 1), different time stages of pregnancy, and different presentations of the foetus.

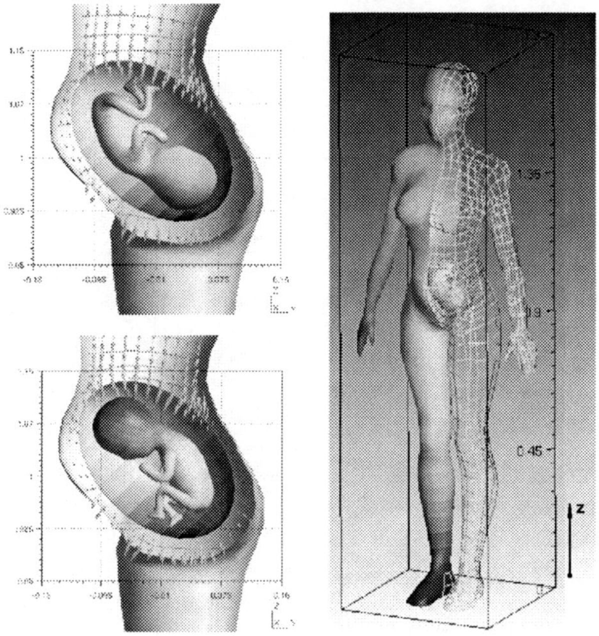

Figure 1: 3D view of the model at 26 weeks of pregnancy in cephalic presentation.

Table 1: Conductivity scenarios.

Scenario	[S/m]	Week 8	Week 13	Week 26	Week 38
	σ_f	0.23	0.23	0.23	0.23
1	σ_{AF}	1.28	1.28	1.27	1.10
	σ_m	0.20	0.20	0.20	0.20
	σ_f	0.996	0.996	0.574	0.574
2	σ_{AF}	1.70	1.70	1.64	1.64
	σ_m	0.52	0.52	0.52	0.52
	σ_f	0.732	0.732	0.396	0.396
3	σ_{AF}	1.70	1.70	1.64	1.64
	σ_m	0.17	0.17	0.17	0.17

4 Numerical implementation

The maternal body is placed in an open environment, standing barefoot on a perfectly conductive infinite flat surface at $z = 0$, at ground level ($\varphi = 0$.) This represents the worst case scenario for open environments in which currents throughout the body are expected to be maximum.

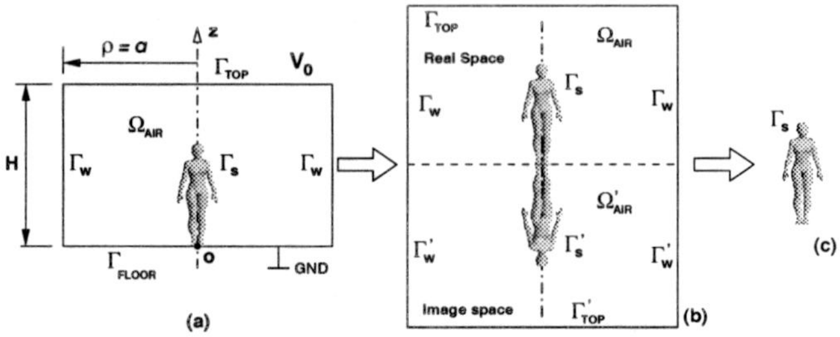

Figure 2: Simplification of the conceptual model. (a) Original. (b) Floor discretisation is avoided by reflecting the problem. (c) Top and lateral wall discretisation is replaced with asymptotic analytical integrations.

The pregnant woman is exposed to a reference field oriented in z direction, with asymptotic value $E_0\hat{\mathbf{z}}$ when $z \to \infty$, as shown in Figure 2. These conditions are recreated by fixing an equi-potential plane $\varphi = V_0$ at $z = H$, where H is sufficiently larger than the height of the woman, and then scaling up the results by a factor $\theta = H/V_0 \times E_0$, in order to translate the results into a particular magnitude of incident field E_0. The assumptions involved have helped to elaborate a particular BEM implementation which reduces dramatically the number of degrees of freedom in the calculation in comparison to the direct BEM application. Thus, the BEM for this work considers a symmetry plane at the ground level, lateral boundary walls located at infinity, and constant potential at a height h enough far away from the model. With these assumptions, only the interfaces between different tissues in the human body and the outer skin in contact with the air around need to be discretised. In this way the mesh discretisation of the external domain including soil, lateral walls and ceiling were eliminated. Figure 2 illustrates the conceptual model developed under these assumptions. The integrals on lateral, top and bottom surfaces of the bounding box in the model were replaced by an equivalent source term: $-zV_0/H$, added to the right hand side member of (1), where z is the vertical coordinate of the source point $\mathbf{x}_i$, V_0 is the potential imposed on the top surface, and H the height of the model. The results in this paper were obtained by adopting $V_0 = 1$ V and $H = 5$ m (i.e. to translate the results to the case of $E_0 = 10$ kV/m, a scaled factor $\theta = 5 \times 10^7$ for $\mathbf{j}$ is adopted in order to obtain $\mathbf{j}$ in mA/m^2).

The model consists of four sub-domains, Ω_{AIR}, Ω_{BODY}, Ω_{AF} and Ω_{F}, namely air, body, AF and foetus, respectively. The air is bounded by Γ_{TOP}, Γ_{W}, Γ_{FLOOR} and Γ_S; while the body is bounded by Γ_{S} and $\Gamma_{ti}, i = 1, \ldots, N_t$, where Γ_{ti} represents the surface enclosing the different internal organs embedded in the homogeneous body, and N_t is the number of organs. Part of Γ_S is in contact to the ground Γ_{SF}, while the rest is in contact to the air Γ_{SA}.

Table 2 summarises the relationship between sub-domains, surfaces and boundary conditions in the model. The complete problem involves the air (considered as an external problem), and the body with its internal organs and the foetus. In order to reduce the computational burden, the BEM solving approach is split into

Table 2: Connectivity between sub-domains and surfaces, and boundary conditions. "×" symbol indicates an unknown of the problem while "-" symbol indicates that the corresponding surface is not related to the sub-domain.

	Ω_{AIR}		Ω_{BODY}		Ω_{AF}		Ω_{F}	
	φ	E_n	φ	E_n	φ	E_n	φ	E_n
Γ_{TOP}	V_0	×	-	-	-	-	-	-
Γ_{W}	$\frac{zV_0}{H}$	0	-	-	-	-	-	-
Γ_{FLOOR}	0	×	-	-	-	-	-	-
Γ_{SA}	×	×	×	×	-	-	-	-
Γ_{SF}	-	-	0	×	-	-	-	-
Γ_{T1}	-	-	×	×	×	×	-	-
Γ_{T2}	-	-	-	-	×	×	×	×

two stages: external and internal. In the external stage the air coupled to the homogeneous body is solved, obtaining as a result the electric field and potential in the skin, which are imposed as input boundary conditions in the second stage in order to solve the interior problem of the human body, consisting of several sub-domains. The inclusion of the body without internal tissues does not introduce significant errors in the results on the skin.

5 Results and discussion

Figure 3 shows a lateral view of the sliced model of the pregnant woman. The direction of the electric field in the maternal tissues is shown with black arrows. The iso-lines represent the electric potential field. Figure 4 shows a 3D view of the sliced model of the pregnant woman including some of the results obtained for the potential and electric field. The model is partially sliced with clipping planes in order to visualise the interior results. There are seven colorbars on the left hand side of the figure which show the correspondence between the colormap and the numerical scale in each case. All results correspond to the case of exposure to 1/5 V/m. "U_body0" corresponds to the potential observed inside the model in the sagittal plane, excluding the limbs. "U_Foetus" refers to the potential measured in the skin of the foetus, the values range from 1.41 μV to 1.34 μV. "U_Skin" corresponds to the potential measured in the maternal skin, i.e. the interface between air and body. The maximum value $\sim 2\ \mu$V is obtained near the head, while the minimum ($\sim O(10^{-3}\ \mu\mathrm{V})$) is in the nearest part of the skin contacting the soil (feet). "E_body" and "E_body0" correspond to the vector field plot representing the electric field in the internal part of the maternal body. The former, ranging from 0.74 μV/m to $\sim$ 0.02 μV/m, refers to the coronal sectioning plane including arms and legs, while the latter, ranging from $\sim$ 2 μV/m to $\sim$ 0.02 μV/m refers to the sagittal planes excluding limbs. Finally, "U_Uterus" and "U_Body" indicate the potential in the uterus and body surfaces, respectively. It can be observed how the uterus tends to concentrate the field lines. This is because of its higher conductivity with respect to the maternal tissue. Figure 4(right) illustrates the observation line that goes along the spine of the foetus, where the current density has been measured. The summary of current density results ($j = | - \sigma_f \nabla\varphi|$), for an incident electric field of 10 kV/m, obtained for 24 models involving cephalic and breech presentations, conductivity scenarios 1, 2, and 3 and four gestational periods can be observed in Figure 5. shows the mean, maximum and minimum values of current density computed in the foetus at different weeks of pregnancy for different conductivity scenarios and foetus presentations. The trend is that the maximum current appears at the 8th gestational week, then it decreases progressively as the foetus develops. This decrease can be explained as a consequence of the two following factors. First, the foetus and AF conductivity decrease with age. Second, as the foetus grows, he tends to adopt a *vertex presentation* (extremities drawn in towards the centre of the chest and head tucked down to the chest), hence his external surface is smoother and the cross sectional area becomes more regular.

In all conductivity scenarios the current density in breech presentation tends to be higher than in the cephalic. This effect is less pronounced in the last stage of pregnancy (38 week).

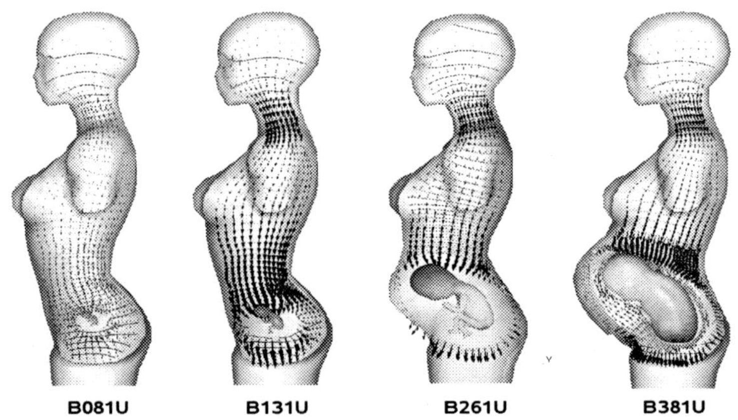

Figure 3: Lateral view of BEM models at 8, 13, 26 and 38 weeks (breech presentation).

Table 3: Dosimetry analysis.

	j_n [mA/m^2]			E_0 [kV/m] field restriction		
Scenario	1	2	3	1	2	3
Maximum	7.4	10.7	17	2.70	1.87	1.18
Typical	3.2	5.8	7.5	6.25	3.45	2.67

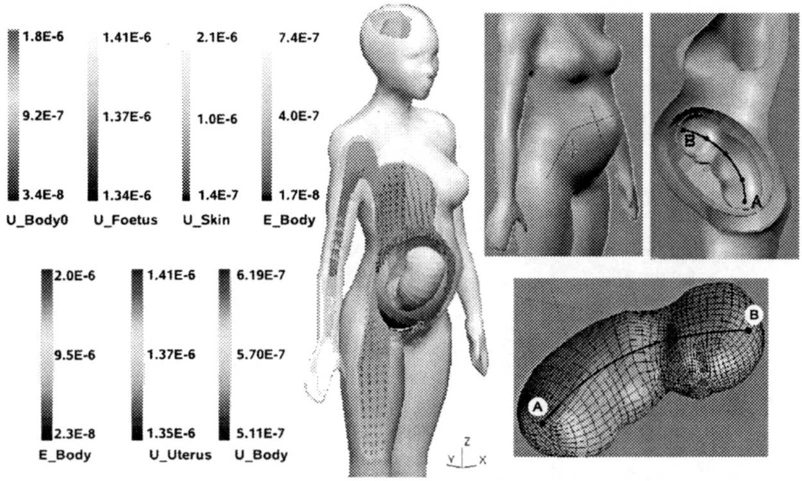

Figure 4: BEM model and exposure results (left). Observation line along the spine of the foetus.

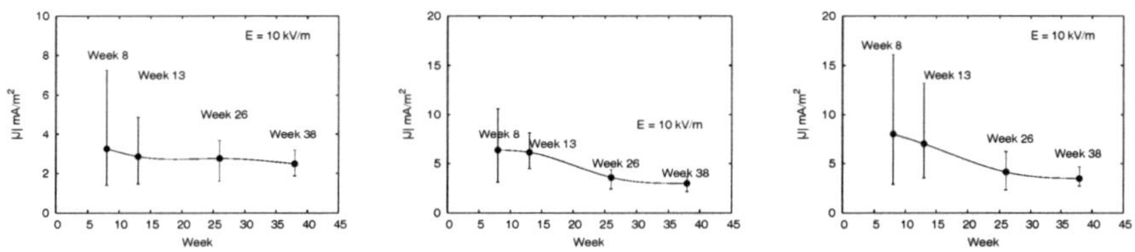

Figure 5: Mean current density calculated in the foetus at weeks: 8, 13, 26 and 38 of pregnancy for scenario 1 (left), 2 (middle), and 3 (right).

6 Summary and conclusions

The BEM has been successfully applied in order to calculate induced currents, potentials and electric fields in a three dimensional anatomical model of a pregnant woman exposed to a vertically incident ELF electromagnetic field. For fixed exposure, the maximum value of current density in the foetus occurs during the 8th week. The maximum current density obtained in the foetus for an incident external field of 10 kV/m is 7.4 mA/m^2. On the other hand, the restriction recommended for public exposure by ICNRP [12] is 2 mA/m^2. Then, this restriction translates into a maximum external field $E_1 = 2.7$ kV/m. Table 3 generalises this analogy for the three scenarios studied by showing the equivalent electric fields restrictions. The first three columns indicate the current densities obtained in the foetus at week 8 for an incident electric field of 10 kV/m in the three different scenarios. The last three columns indicate the external electric field that should be applied in order to measure a current density of 2 mA/m^2 in the foetus. The first row corresponds to the maximum values while the second to the average ones. According to these results, the mean values give a restriction for the electric field of approximately 3 kV/m based on the current circulating in the foetus, while the current restriction suggests approximately 50 kV/m, if it is based only on the maternal brain.

References

[1] P.J. Dimbylow. Development of pregnant female, hybrid voxel mathematical models and their application to the dosimetry of applied magnetic and electric fields at 50 hz. *Phys. Med. Biol.*, 51:2383–94, 2006.

[2] C. Brebbia, J. Telles, and L. Wrobel. *Boundary Elements Techniques*. Splinger-Verlag, Berlin, Heidelberg, New York and Tokio, 1984.

[3] C. Brebbia and J.Dominguez. *Boundary Elements, an Introductory Course*. Computational Mechanics Publications. McGraw-Hill, New York, Colorado, San Francisco, Mexico, Toronto, 2nd edition, 1992.

[4] C. Gonzalez, A. Peratta, and D. Poljak. Boundary element modeling of the realistic human body exposed to extremely low frequency (elf) electric fields: Computational and geometrical aspects. *IEEE Trans. on Electromagnetic Compatibility*, 49(1):153–62, 2007.

[5] A. Peratta, C.Gonzalez, and D. Poljak. Geometrical aspects of 3d human body exposed to extremely low frequency electromagnetic fields. In *14th Int.Conf on Software, Telecomm and Comp. Networks. IEEE*, Split CROATIA, 2006.

[6] A. Peratta, C. Gonzalez, and D. Poljak. Current density induced in the human body due to power distributions lines using the BEM. *J. Comm. Soft. Syst.*, 3(1):11–16, 2007.

[7] A. Peratta. 3D Low frequency electromagnetic modelling of the human eye with boundary elements: Application to conductive keratoplasty. *Engineering Analysis with Boundary Elements*, 2008. To appear in 2008.

[8] S.Gabriel, R.W. Lau, and C. Gabriel. The dielectric properties of biological tissues: Ii. measurements in the frequency range 10 hz to 20 ghz phys. *Phys. Med. Biol.*, 41:2251–69, 1996.

[9] MJT FitzGerald MJT and M. FitGerald. *Human Embryology*. W. B Sauders Company, London, 1994.

[10] UNSW Embriology - Dr.Mark Hill. Website.

[11] ICRP 2002. Basic anatomical and physiological data for use in radiological protection: reference values. *ISNN 0146-6453*, 89, 2002.

[12] ICNIRP. Guidelines for limiting exposure to time-varying electric, magnetic, and electromagnetic fields (up to 300 ghz). *Health Physics.*, 74(4):494–522, 1998.

Trefftz analysis for temperature rise on human skin exposed to electromagnetic waves

Y. Hirayama & E. Kita
Graduate School of Information Sciences, Nagoya University, Nagoya 464-8301, Japan

Abstract

This paper describes the application of Trefftz method to the temperature rise in a human skin exposed to a radiation from a cellular phone. A governing equation is given as Poisson equation. An inhomogeneous term of the equation is approximated with a polynomial function in Cartesian coordinates. The use of the approximated term transforms the original boundary-value problem to that governed with a homogeneous differential equation. The transformed problem can be solved by the traditional Trefftz formulation. Firstly, the present method is applied to a simple numerical example in order to confirm the formulation. The temperature rise in a skin exposed to a radiation is considered as a second example.
Keywords: Trefftz method, poisson equation, polynomial function.

1 Introduction

There has been an increasing public concern regarding the possible health effects of human exposure to an electromagnetic radiation. In Japan, there are four cellular phone carrier companies; NTT Docomo, au, Softbank (vodafone) and Willcome. They provide different types of cellular phone services. NTT Docomo, au and Softbank (vodafone) provides Code Division Multiple Access (CDMA)-type phones and Willcome Personal Handy-phone System (PHS)-type ones. The CDMA-type phones are so-called 3G-cellular phones, which are used widely around the world. The PHS-type phones operate in the 1880–1930 MHz frequency band, which are used mainly in Japan, China, Taiwan and some other Asian countries. Since carrier companies operate different services, their health effects are also dissimilar. In this paper, we will focus on the temperature rise in a human skin exposed to the radiation. The hazardous electromagnetic field levels can be quantified analyzing the thermal response of the human body exposed to the radiation [1, 2].

WIT Transactions on Modelling and Simulation, Vol 47, © 2008 WIT Press
www.witpress.com, ISSN 1743-355X (on-line)
doi:10.2495/BE080101

The temperature rise in a human skin exposed to the radiation can be defined as a boundary-value problem of a Poisson equation. Original boundary element methods are not adequate for solving an inhomogeneous differential equation such as Poisson equation. Since there exists a domain integral term derived from an inhomogeneous term of the equation in the integral equation, domain discretization is necessary. For transforming the domain integral due to a boundary one, several formulations have been presented; dual reciprocity method [3, 4], multiple reciprocity method [5, 6], radial bases function approximation [7–11], polynomial function approximation method [12], boundary point interpolation method [13], and so on.

In this paper, Trefftz method based on the collocation formulation is applied to the problem [14–16]. Since an integral equation is not necessary, the above-mentioned formulations can not be applied directly for the formulation. So, we will use the polynomial function approximation method of the particular solution of the problem. The problem to be solved is governed with a Poisson equation. An inhomogeneous term of the equation is approximated with a polynomial function in Cartesian coordinates. The approximated expression of the particular solution can be derived easily. The use of the approximated particular solution transforms the boundary value problem of Poisson equation to that of Laplace equation. The boundary value problem governed with Poisson equation can be solved easily according to the traditional Trefftz formulation.

The remaining of the paper is organized as follows. In section 2, the formulation of the problem to be solved is described. In section 3, the present method is applied to some numerical examples. The section 4 summarizes the conclusions.

2 Formulation

2.1 Governing Equation and Boundary Condition [1, 2]

For harmonically varying electromagnetic field, the temperature rise on the human skin can be calculated from the bio-heat equation

$$C\rho\frac{\partial u}{\partial t} = K\nabla^2 u + Q_{em} - B(u - u_b), \tag{1}$$

where u is the temperature of the tissue, u_b is the temperature of the blood, K is the thermal conductivity of the tissue, C is the heat capacity of the tissue, B is the term associated with blood flow, and Q_{em} is the electromagnetic power deposition.

The electromagnetic power deposition Q_{em} is given as

$$Q_{em} = \rho \cdot SAR \tag{2}$$

where ρ and SAR denote a tissue density and specific absorption rate (SAR), respectively. For harmonically varying electromagnetic field, the SAR is defined as

$$SAR = \frac{\sigma}{2\rho}|E|^2 \tag{3}$$

where $|E|$ is the peak value of electromagnetic-field, and σ is the conductivity of the tissue.

In addition, the boundary condition for equation (1) is given by

$$u = \bar{u}, \tag{4}$$

$$q \equiv \frac{\partial u}{\partial n} = \bar{q}, \tag{5}$$

and, on the skin surface,

$$h \cdot (u - u_a) = -Kq \tag{6}$$

where H and u_a are the convection coefficient and temperature of the air, respectively.

At the thermal steady state, $\partial u/\partial t = 0$ and therefore, equation (1) is reduced to the

$$\nabla^2 u + b = 0, \tag{7}$$

where

$$b = \frac{1}{K}[\rho \cdot SAR - B(u - u_a)]. \tag{8}$$

The boundary conditions (4) to (6) are rewritten as

$$\alpha u + \beta q = \gamma, \tag{9}$$

where α, β and γ denote the parameters derived from the boundary conditions.

2.2 Trefftz Formulation [17]

Now, we consider the boundary value problem defined by equations (7) and (9). An inhomogeneous term (8) is approximated with a polynomial function in Cartesian coordinates;

$$b = \boldsymbol{c}^T \boldsymbol{r}. \tag{10}$$

In this study, the fifth-order polynomial is adopted for the function $\boldsymbol{r}$. Therefore, $\boldsymbol{c}$ and $\boldsymbol{r}$ are defined respectively as follows

$$\boldsymbol{c}^T = \{c_1, c_2, \cdots, c_{21}\}, \tag{11}$$

$$\begin{aligned} \boldsymbol{r}^T &= \{r_1, r_2, r_3, \cdots, r_{21}\} \\ &= \{1, x, y, x^2, xy, y^2, x^3, x^2y, xy^2, y^3, y^3, x^4, x^3y, \\ &\quad x^2y^2, xy^3, y^4, x^5, x^4y, x^3y^2, x^2y^3, xy^4, y^5\}. \end{aligned} \tag{12}$$

The use of the equation (10) transforms the original governing equation into

$$\nabla^2 u + \boldsymbol{c}^T \boldsymbol{r} = 0. \tag{13}$$

Since the term r_i is a polynomial function, the related particular solution u_i^p can be determined easily. u_i^p satisfies the equation

$$\nabla^2 u_i^p + r_i = 0. \tag{14}$$

In the Trefftz method, the homogeneous solution of the governing equation u^h is approximated with the superposition of the related T-complete function u_i^* [18]. The unknown function u is approximated with the T-complete function u_i^* and the particular solution u_i^p as follows.

$$u = u^h + \boldsymbol{c}^T \boldsymbol{u}^p = \boldsymbol{a}^T \boldsymbol{u}^* + \boldsymbol{c}^T \boldsymbol{u}^p, \tag{15}$$

where $\boldsymbol{a}$ denotes the unknown parameter vector for approximating the homogeneous solution. Besides, u_i^* and u_i^p satisfy the equations

$$\begin{aligned} \nabla^2 u_i^* &= 0, \\ \nabla^2 u_i^p + r_i &= 0. \end{aligned}$$

Equation (15) satisfies Eq. (7) but dose not satisfy Eq. (9). Substituting Eq. (15) into (9) leads to residual expressions:

$$\begin{aligned} R &= \alpha u + \beta q - \gamma \\ &= \alpha(\boldsymbol{a}^T \boldsymbol{u}^* + \boldsymbol{c}^T \boldsymbol{u}^p) + \beta(\boldsymbol{a}^T \boldsymbol{q}^* + \boldsymbol{c}^T \boldsymbol{q}^p) - \gamma \\ &= (\alpha \boldsymbol{u}^* + \beta \boldsymbol{q}^*)^T \boldsymbol{a} + (\alpha \boldsymbol{u}^p + \beta \boldsymbol{q}^p)^T \boldsymbol{c} - \gamma \end{aligned} \tag{16}$$

where

$$\left. \begin{aligned} q_i^* &= \frac{\partial u_i^*}{\partial \boldsymbol{n}} \\ q_i^p &= \frac{\partial u_i^p}{\partial \boldsymbol{n}} \end{aligned} \right\}. \tag{17}$$

Satisfying the residual equations at the boundary point P_m by means of the collocation formulation, we have

$$\boldsymbol{K}\boldsymbol{a} = \boldsymbol{f} - \boldsymbol{B}\boldsymbol{c}, \tag{18}$$

In the matrix $\boldsymbol{K}$, the total numbers of the rows and the columns are equal to the total number of the boundary collocation points and the T-complete functions, respectively. Therefore, we shall take more collocation points than the T-complete functions; i.e., $M > N$ and Equation (18) is solved by using the singular value decomposition method of the LAPACK software [19].

2.3 Determination of parameter *c*

The unknown parameter vector $\boldsymbol{c}$ in Eq. (18) is determined by using the iterative process. Equation (10) held at the iteration steps (k) and $(k+1)$ are

$$b^{(k+1)} = \boldsymbol{r}^T \boldsymbol{c}^{(k+1)},$$
$$b^{(k)} = \boldsymbol{r}^T \boldsymbol{c}^{(k)}.$$

Subtracting both sides of the above equations leads to

$$\begin{aligned} b^{(k+1)} - b^{(k)} &= \boldsymbol{r}^T (\boldsymbol{c}^{(k+1)} - \boldsymbol{c}^{(k)}), \\ \Delta b &= \boldsymbol{r}^T \Delta \boldsymbol{c}, \end{aligned} \tag{19}$$

where the superscript (k) denotes the number of iteration.

The collocation points, which are referred as "the computing point", are placed on the boundary and within the domain. Holding Equation (19) on the computing points and arranging them in the matrix form, we have

$$\boldsymbol{D} \Delta \boldsymbol{c} = \boldsymbol{f}, \tag{20}$$

where $\boldsymbol{D}$ and $\boldsymbol{f}$ denote the coefficient matrix and vector, respectively. Equation (20) is solved for $\Delta \boldsymbol{c}$ with the singular value decomposition of LAPACK software [19]. The parameter $\boldsymbol{c}$ is updated with

$$\boldsymbol{c}^{(k+1)} = \boldsymbol{c}^{(k)} + \Delta \boldsymbol{c}. \tag{21}$$

The convergence criterion is defined as

$$\eta \equiv \frac{1}{M_c} \sum_{i=1}^{M_c} |\Delta b(Q_i)| < \eta_c \tag{22}$$

where M_c and η_c denote the total number of the computing points and the positive constant specified by a user, respectively. Q_i denotes the computing points placed on the boundary and within the domain.

3 Numerical examples

The object under consideration is a square region of $-1 \leq x \leq 1, -1 \leq y \leq 1$ (Fig. 1). Figure 1 shows the part of a human skin. The right segment is exposed to the air and the other segments are in the body. The physical parameters are given as follows; $\rho = 1010[\mathrm{kg/m^3}]$, $K = 0.50[\mathrm{W/m} \cdot {}^\circ]$, $B = 8650[\mathrm{W/m^3}{}^\circ]$, $u_b = 37.0[{}^\circ]$, $h = 10.5[\mathrm{W/m^3}{}^\circ]$, and $u_a = 27.0[{}^\circ]$.

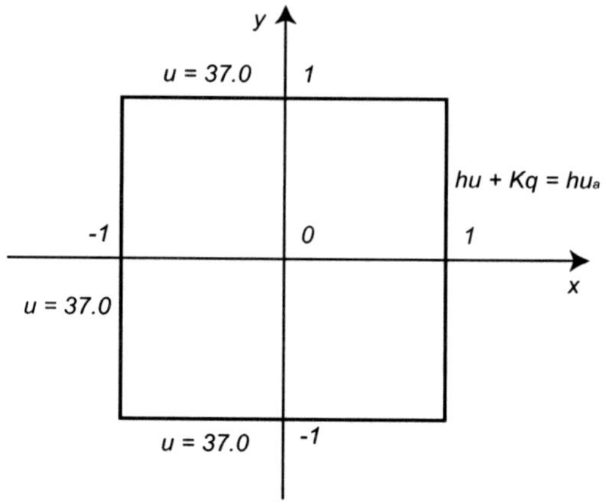

Figure 1: Numerical example.

Table 1: Numerical results.

Model	SAR	Max. Temp. Rise [°]
Docomo SO702i	1.35	0.000409
Nokia 702NK2	0.83	0.000251
au Talby	0.499	0.000151
Willcom WA007SH	0.021	0.000006

The parameter SAR[W/kg] depends on cellular phones. We will consider four cellular phones; Docomo (SO702i), au(Talby), Softbank(Nokia 702NK2), and Willcom(WA007SH), which are popular in Japan. The boundary conditions are given as

$$\left.\begin{aligned} hu + Kq &= hu_a \quad (x = 1) \\ u &= 37.0 \quad (y = 1) \\ u &= 37.0 \quad (x = -1) \\ u &= 37.0 \quad (y = -1) \end{aligned}\right\} \tag{23}$$

Numerical results are shown in Table 1. Distributions of the temperature rise are shown in Figs. 2 to 5. The temperature rise increases according to the increase of the SAR value. They are 3G-type cellular phones, except for Willcom(WA007SH). However, from the view-point of the rise in the human skin, we notice that Willcom(WA007SH) is the most gentle and that the au(Talby) is gentler than the other 3G-type cellular phones.

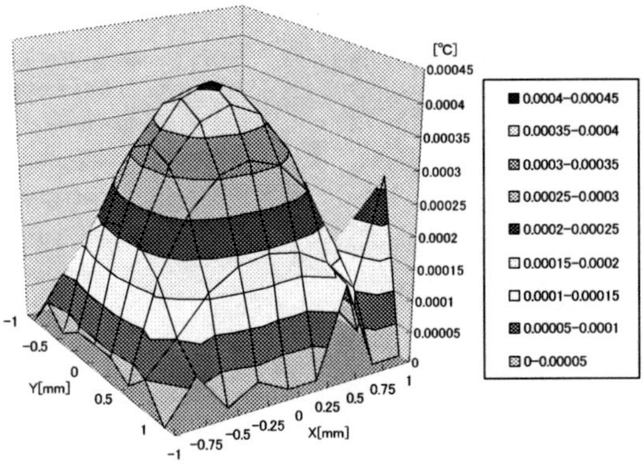

Figure 2: Docomo (SO702i).

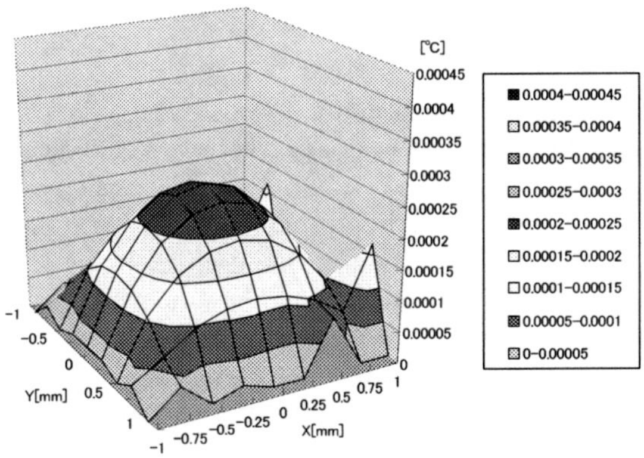

Figure 3: Softbank (Nokia 702NK2).

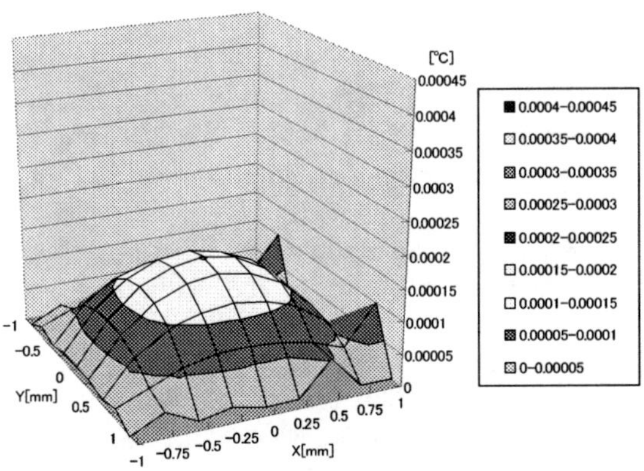

Figure 4: au (Talby).

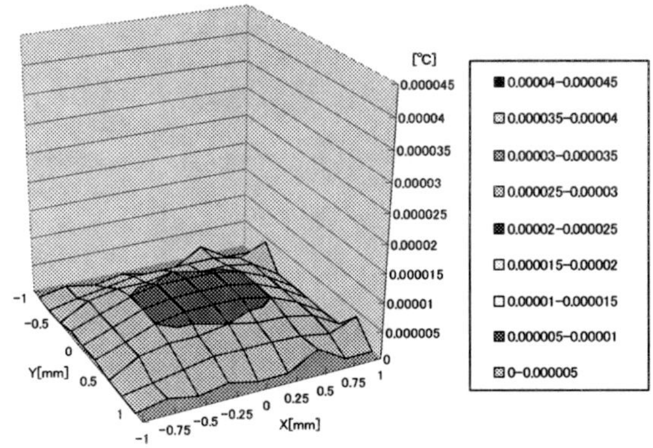

Figure 5: Willcom (WA007SH).

4 Conclusions

This paper describes the application of the Trefftz method to the temperature rise analysis in a human skip exposed to a radiation from a cellular phone. The governing equation is given as Poisson equation. An inhomogeneous term of the equation is approximated with a polynomial function in Cartesian coordinates. The use of the approximated term transforms the original boundary-value problem to that governed with a homogeneous differential equation. The transformed problem can be solved by the traditional Trefftz formulation. We compare four cellular phones from the view-point of the rise in the human skin; NTT Docomo (SO702i), au(Talby), Softbank(Nokia 702NK2), and Willcom(WA007SH), which are popular in Japan. They are 3G-type cellular phones, except for Willcom(WA007SH). We notice that Willcom(WA007SH) is the most gentle and that the au(Talby) is gentler than the other 3G-type cellular phones.

References

[1] A. Hirata and T. Shiozawa. Correlation of maximum temperature increase and peak SAR in the human head due to handset antennas. *IEEE, Transactions on Microwave Theory and Techniques*, Vol. 51, No. 7, pp. 1834–1841, 1999.

[2] D. Poljak, N. Kovac, T. Samardzioska, A. Peratta, and C. A. Brebbia. Temperature rise in the human body exposed to radiation from base station antennas. In C. A. Brebbia, editor, *Boundary Elements XXVI (Proc. 26th World Conf. on BEM, Bologna, Italy, 2004)*, pp. 381–390. WIT Press, 2004.

[3] T. W. Partridge, C. A. Brebbia, and L. C. Wrobel. *The Dual Reciprocity Boundary Element Method*. Comp. Mech. Pub. / Springer Verlag, 1992.

[4] T. W. Partridge. Towards criteria for selecting approximation functions in the dual reciprocity method. *Engineering Analysis with Boundary Elements*, Vol. 24, No. 7, pp. 519–529, 2000.

[5] A. J. Nowak and A. C. Neves. *The Multiple Reciprocity Boundary Element Method*. Comp. Mech. Pub. / Springer Verlag, 1994.

[6] A. J. Nowak. Application of the multiple reciprocity BEM to nonlinear potential problems. *Engineering Analysis with Boundary Elements*, Vol. 18, pp. 323–332, 1995.

[7] Z. Liu and J. G. Korvnik. Accurate solving the Poisson equation by combining multiscale radial basis functions and Gaussian quadrature. In C. A. Brebbia, A. Tadeu, and V. Popov, editors, *Boundary Elements XXIV (Proc. 24th World Conf. on BEM, Sintra, Portugal, 2002)*, pp. 97–104. Comp. Mech. Pub., 2002.

[8] W. Chen. Some recent advances on the RBF. In C. A. Brebbia, A. Tadeu, and V. Popov, editors, *Boundary Elements XXIV (Proc. 24th World Conf. on BEM, Sintra, Portugal, 2002)*, pp. 125–134. Comp. Mech. Pub., 2002.

[9] V. M. A. Leitao and C. M. Tiago. The use of radial basis functions for one-dimensional structural analysis problmes. In C. A. Brebbia, A. Tadeu, and V. Popov, editors, *Boundary Elements XXIV (Proc. 24th World Conf. on BEM, Sintra, Portugal, 2002)*, pp. 165–179. Comp. Mech. Pub., 2002.

[10] B. Sarler, J. Perko, and C.S. Chen. Radial basis function collocation method solution of natural convection in porous media. *International Journal of Numerical Methods for Heat and Fluid Flow*, Vol. 14, No. 2, pp. 187–212, 2004.

[11] M.S. Ingber, C.S. Chen, and J.A. Tanski. A mesh free approach using radial basis functions and parallel domain decomposition for solving three dimensional diffusion equations. *International Journal for Numerical Methods in Engineering*, Vol. 60, pp. 2183–2201, 2004.

[12] Xu S.Q. and Kamiya N. A formulation and solution for boundary element analysis of inhomogeneous-nonlinear problem; the case involving derivatives of unknown function. *Engineering Analysis with Boundary Elements*, Vol. 23, No. 5/6, p. 391, 1999.

[13] G. R. Liu and Y. T. Gu. Boundary meshfree methods basd on the boundary point interpolation methods. In C. A. Brebbia, A. Tadeu, and V. Popov, editors, *Boundary Elements XXIV (Proc. 24th World Conf. on BEM, Sintra, Portugal, 2002)*, pp. 57–66. Comp. Mech. Pub., 2002.

[14] E. Trefftz. Ein Gegenstück zum Ritzschen Verfahren. *Proc. 2nd Int. Cong. Appl. Mech., Zurich*, pp. 131–137, 1926.

[15] Y. K. Cheung, W. G. Jin, and O. C. Zienkiewicz. Direct solution procedure for solution of harmonic problems using complete, non-singular, Trefftz functions. *Communications in Applied Numerical Methods*, Vol. 5, pp. 159–169, 1989.

[16] M. Karas and A. P. Zielinski. Application of Trefftz complete functional system to stress analysis in helical spring with an arbitrary wire cross-section. *Strojnicky Casopis*, Vol. 49, pp. 426–437, 1998.

[17] E. Kita, Y. Ikeda, and N. Kamiya. Trefftz solution for boundary value problem of three-dimensional poisson equation. *Engineering Analysis with Boundary Elements*, Vol. 29, pp. 383–390, 2005.

[18] I. Herrera. Theory of connectivity: A systematic formulation of boundary element methods. In C. A. Brebbia, editor, *New Developments in Boundary Element Methods (Proc. 2nd Int. Seminar on Recent Advances in BEM, Southampton, England, 1980)*, pp. 45–58. Pentech Press, 1980.

[19] E. Anderson, Z. Bai, C. Bischof, J. Demmel, J. Dongarra, J. Du Croz, A. Greenbaum, S. Hammarling, A. McKenney, S. Ostrouchov, and D. Sorensen. *LAPACK User's Manual*. SIAM, 2.nd edition, 1995.

Modelling of protective potential distribution in a cathodic protection system using a coupled BEM/FEM method

A. Muharemović[1], H. Zildžo[1] & E. Letić[2]
[1]*Department of Electrical Engineering, University of Sarajevo, Bosnia-Herzegovina*
[2]*Energoinvest d.d. Sarajevo, Bosnia-Herzegovina*

Abstract

Design of any cathodic protection (CP) system requires one to determine current density which meets predefined criteria, and to ensure as uniform a current density distribution as possible over the structure surface. Nowadays the only reliable way to determine current and potential distribution is employment of numerical techniques. In this paper, the optimisation of a passive cathodic protection system by using coupled BEM/FEM is presented. FEM is appropriate for solving completely bounded domains, and it is used for discretization of anodes and cathodes. BEM is applied for discretization of electrolyte, i.e. infinite domain. In the coupled BEM/FEM the soil domain is linked to the steel domain through the current density (boundary condition) generated by the kinetics of the corrosion reactions at the steel surface. Verification of the obtained numerical result is made by comparing it with measured results. Good agreement between simulation and measurements has been found.

Keywords: cathodic protection, finite element method, boundary element method, galerkins weighted residuals method, successive underrelaxation method.

1 Introduction

The distribution of protection potentials in systems of cathodic protection (electrostatic field with or without sources) is defined either by the Poisson or Laplace equation, with the usage of corresponding boundary conditions.

www.witpress.com, ISSN 1743-355X (on-line)
doi:10.2495/BE080111

Boundary conditions in the cathodic protection system, in most cases, are not linear because the electrochemical reactions do not give linear functional relations between the protection current density and the corresponding potential. This piece will treat nonlinear functional relations V = f (σ) and the case of a homogeneous electrolyte, i.e. electrolyte with the same average value of soil resistivity [2].

On the figure 1 the basic model of cathodic protection is given. The galvanic anode is made of cast zinc. Positive ions from the anode move through the electrolyte towards the object that is being protected, while electrons move in the opposite direction. DP (drain point) represents the point where the cable installation connects with the object being protected – the pipeline.

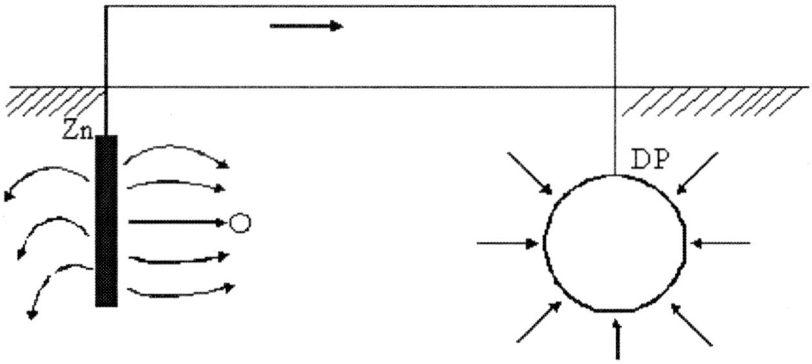

Figure 1: Cathodic protection system with galvanic anodes.

The authors have developed software for modeling and optimization of cathodic protection systems with galvanic anodes. Idea for this project is based on the works in field of cathodic protection that Bosnian company Energoinvest carried out in Libya in period 2003-2005 year. Owner and investor of the works was Great Man-made River Authority (GMMRA). The contractual obligation was to protect cathodically a pipeline in total length of 204 km, and with internal diameter of 4 meters. All the measured results before and after CP system installation were available and these results enable the authors to make software validation i.e. comparison measurements and numerical calculation outcome.

2 CP system design

The system was designed to provide a 250 mV IR free potential shift criteria from the natural potential, by using a 1mA/sq.m current density at the soil resistivity of 3000 ohm-cm or less, and at higher soil resistivity area, a lower current is expected to be required to provide the required protection [1]. The criterion of 250 mV IR free potential shift criteria was selected based on the above ground surface measurement to ensure that the criteria of 100 mV IR free potential shift could be achieved and guaranteed over circumference of the pipeline. The anode assemblies have been installed in vertically drilled holes on both sides of the pipeline at normally 5.5 meters distance from the pipeline's centre line. The anode spacing might vary from groundbed to groundbed, but the average spacing is 13.5 meters for cast zinc (CZ) anode type.

3 Mathematical model

The Finite Element Method (FEM) is appropriate for the domains with several media that have finite boundaries. The Boundary Element Method (BEM) is suitable for calculation of the domains which contain only one medium and boundaries that can be finite or infinite. Let's consider the basic principle of the Galerkin's weighted residuals procedure used in the FEM, direct BEM, and iterative sequential Dirichlet-Neumann's hybrid BEM/FEM method. Governing Laplace partial differential equation (PDE) is solved by using the BEM. Integral equation of the steady state current field by using the FEM.

The potentials on the BEM/FEM boundary are calculated by applying the successive underrelaxation iterative method [10].

3.1 The Finite Element Method (FEM)

In the FEM, the domain of the considered physical system is divided into finite number of elements of certain geometry called finite elements using so-called procedure discretization of continuum [4–6]. The governing Laplace PDE of the steady state current field is

$$\frac{\partial}{\partial x}\left(\sigma\frac{\partial\varphi}{\partial x}\right)+\frac{\partial}{\partial y}\left(\sigma\frac{\partial\varphi}{\partial y}\right)+\frac{\partial}{\partial y}\left(\sigma\frac{\partial\varphi}{\partial y}\right)=0 \tag{1}$$

Applying the Galerkin's weighted residuals procedure the solution of electric potentials distribution can be written in the following form:

$$[H]^{FEM}\cdot\{\varphi\}^{FEM}=\{Q\}^{FEM} \tag{2}$$

where

$[H]^{FEM}$ - two-dimensional matrix of coefficients where the common term is given by:

$$h_{ij}^{FEM}=\sum_{e=1}^{n_e}\left[\sigma\int_{V^e}\left(\frac{\partial N_i^e}{\partial x}\frac{\partial N_j^e}{\partial x}+\frac{\partial N_i^e}{\partial y}\frac{\partial N_j^e}{\partial y}+\frac{\partial N_i^e}{\partial z}\frac{\partial N_j^e}{\partial z}\right)dV\right]$$

$$(i=1,2,..,n_f;\ j=1,2,...,n_f) \tag{3}$$

$\{\Phi\}^{FEM}$ - column vector matrix of unknown potentials in the nodes of a finite element. Order of this element is n_f x 1.

$\{Q\}^{FEM}$ - column vector matrix of the free terms that contain Neumann boundary conditions. The common term is given by:

$$q_i^{FEM} = -\sum_{e=1}^{n_e}\left[\sum_{j=1}^{n_f}\left(\int_{S_\Delta^e}\sigma\, N_i^e \cdot N_j^e \cdot \frac{\partial \varphi_j^{FEM}}{\partial n} dS\right)\right] \tag{4}$$

N_i^e - shape functions that are used to approximate unknown potentials function in the following way:

$$\phi^{FEM} = \sum_{j=1}^{n_f} N_j^e \cdot \phi_j^e \tag{5}$$

$\frac{\partial \varphi_j^{FEM}}{\partial n}$ - Neumann boundary condition.

3.2 The direct boundary element method

The mathematical model of the direct BEM is based on Green's symmetrical identity as well as equations of continuity [3]. These equations serve to apply the boundary conditions on the boundaries between different media. Let's consider two different cases of 3D electrostatic field calculation. The first case is when the observed point Q falls within the domain V, and the second case is when it falls on the boundary of media.

Generally, formula for potential calculation within, and out of the domain as well as on the boundary is given by:

$$C(Q)\cdot\varphi(Q) + \int_S T(P,Q)\cdot\varphi(P)\cdot dS_P = \int_S G(P,Q)\cdot\frac{\partial\varphi(P)}{\partial n_P}\cdot dS_P \tag{6}$$

where

$G(P,Q)$ - Green function,

$T(P,Q)$ - derivative of the Green function in the direction of outward normal vector to the boundary surface,

φ i $\frac{\partial\varphi}{\partial n}$ - calculated the potential and normal electric field component on the boundary surface

$C(Q)$ - the constant that can be represented by:

$$C(Q) = \begin{cases} 1 & \text{inside domain V (Poisson formula)} \\ \frac{1}{2} & \text{on smooth boundary in 3D domain} \\ \frac{\gamma_{3-D}}{4\pi} & \text{on non-smooth boundary in 3D domain} \\ 0 & \text{outside of domain V} \end{cases} \tag{7}$$

Applying the collocation point procedure of the weighted residuals method on the equation (6) the solution in the matrix system form can be written:

$$\left[H\right]^{BEM} \cdot \{\varphi\}^{BEM} = \left[G\right]^{BEM} \cdot \left\{\frac{\partial \varphi}{\partial n}\right\}^{BEM} \tag{8}$$

where

$\left[H\right]^{BEM}$ - two-dimensional coefficient matrix with general term defined by:

$$h_{i,j}^{BEM} = \sum_{e=1}^{n_e} \int_S N_j^e \cdot T_{i,j}^e \cdot dS_p + \delta_{i,j} \cdot C_i$$

$$(i=1,2,..,n_e\ j=1,2,...,n_e) \tag{9}$$

$\left[G\right]^{BEM}$ - two-dimensional coefficient matrix with general term defined by:

$$g_{i,j}^{BEM} = \sum_{e=1}^{n_e} \int_S N_j^e \cdot G_{i,j}^e \cdot dS \qquad (i=1,2,..,n_e\ j=1,2,...,n_e) \tag{10}$$

$\{\varphi\}^{BEM}$ i $\left\{\frac{\partial \varphi}{\partial n}\right\}^{BEM}$ – vector column matrix of variables.

The value of variable φ or $\partial\varphi/\partial n_p$ is known in every single node of the boundary elements for the domain boundary with only one media. So by using the equation (8) the calculation of the variable (φ or $\partial\varphi/\partial n_p$) that is not given as a boundary condition (φ or $\partial\varphi/\partial n_p$) is found.

On the boundary between two domains with different media the unknowns are φ or $\partial\varphi/\partial n_p$. In this case separate equation system (6) is formed for each boundary of domain taking into consideration the Dirichlet's and Neumann's boundary conditions. The polarization curve, which describes the relationship between the potential and the current density, indicates the corrosion condition on the pipe surface and is normally used as the boundary condition. It is not an easy task to determine the polarization curve since it strongly depends on a number of phenomena. Furthermore, the polarization curve can also be a function of time and history. One common polarization curve for steel (cathode) can be written in the following form [7]:

$$i = \frac{A_{pore}}{A}\left[10^{\frac{V-\phi-E_{F_e}}{\beta_{F_e}}} - \left(\frac{1}{\left(1-\alpha_{blk}\right)\cdot i_{\lim,O_2}} - 10^{\frac{V-\phi-E_{O_2}}{\beta_{O_2}}}\right)^{-1} - 10^{\frac{-(V-\phi-E_{H_2})}{\beta_{H_2}}}\right] \tag{11}$$

The polarization curve for a galvanic anode would be:

$$i = i_{o_2}\left(10^{\frac{V-\phi-E_{corr}}{\beta_{anode}}} - 1\right) \tag{12}$$

Where V is the potential of the steel, obtained from solution of the inner domain, and Φ is the potential of the soil adjacent to the steel, obtained from solution of the outer domain. The term β_{Fe} represents the Tafel slope for the corrosion reaction and E_{Fe} represents the equilibrium potential for the corrosion reaction [7].

On the boundary between two domains with different media additional equations of continuity for φ or $\partial\varphi/\partial n_p$ are written.

3.3 The hybrid BEM/FEM method

An example of the cathodic protection system calculation using 3D finite elements discretization of the anodes and cathode is shown on the figure 2. Discretization of electrolyte using boundary elements as well as metallic connections between anodes and cathode applying 1D finite elements are also shown on the figure 2.

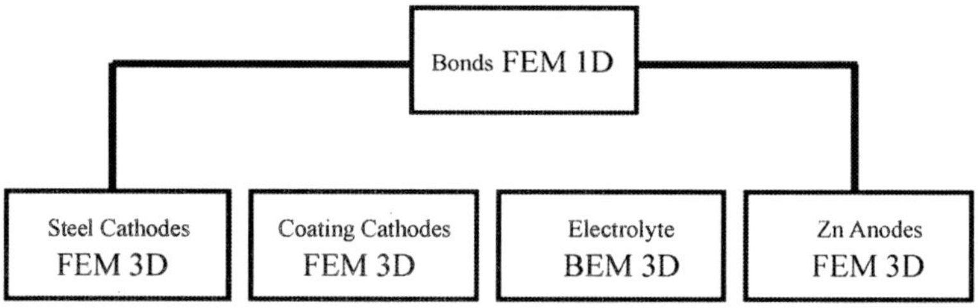

Figure 2: An example of discretization of a cathodic protection system using the coupled BEM/FEM domain.

There are direct and iterative algorithms for linking the finite and boundary element methods. The direct approach requires a system of linear algebraic equations given by

- equation (2) in the FEM domain
- equation (8) in the BEM domain

and the continuity equations for the BEM/FEM boundary.

The significant disadvantage of this approach is the need to solve large system of equations. In order to save the CP memory it is recommended to use some of the following algorithms:

- Robin relaxation coupled algorithm
- Neumann-Neumann coupled algorithm
- Advanced Dirichlet-Neumann algorithm
- Advanced sequential Dirichlet-Neumann algorithm

When using the BEM/FEM iterative method separate calculations for the two separate systems of linear algebraic equations are used, one for the BEM and the other for the FEM domain. The potential distribution or flux, at the BEM/FEM boundary, is then found using the iterative successive underrelaxation method.

3.4 Advanced sequential Dirichlet-Neumann algorithm

In this paper advanced sequential Dirichlet-Neumann BEM/FEM algorithm is used. This algorithm consists of the following steps [10]:

- Partition of the whole domain into BEM and FEM domains
- Set initial potential values on the BEM/FEM boundary

- Start the iterative procedure which lasts until the set convergence condition is fulfilled.

3.4.1 Calculation of the electric field in BEM domain

The Dirichlet or Neumann boundary conditions are given on the boundary of the BEM domain, except on the boundary between the BEM and FEM domains. Therefore, for the matrix system (8) in matrices $[H]^{BEM}$ and $[G]^{BEM}$ the BEM-BEM contributions from "pure" BEM boundary and the BEM/FEM contribution from the mutual BEM/FEM boundary could be written in the following form:

$$[H]^{BEM}\begin{Bmatrix}\{\varphi\}_{n+1}^{BEM-BEM}\\ \{\varphi\}_{n+1}^{BEM-FEM}\end{Bmatrix}=[G]^{BEM}\begin{Bmatrix}\left\{\dfrac{\partial\varphi}{\partial n}\right\}_{n+1}^{BEM-BEM}\\ \left\{\dfrac{\partial\varphi}{\partial n}\right\}_{n+1}^{BEM-FEM}\end{Bmatrix} \tag{13}$$

Also in the system (13) it is necessary to take into account the Dirichlet and Neumann boundary conditions including the potential values at the BEM/FEM boundary from the previous iteration step. Solving the system (13) the normal components of the electric field $\left\{\dfrac{\partial\varphi}{\partial n}\right\}_{n+1}^{BEM-FEM}$ at the BEM/FEM boundary will be found.

3.4.2 Calculation of continuity equation at the BEM/FEM boundary

$$\begin{gathered}\sigma^{FEM}\left\{\frac{\partial\varphi}{\partial n}\right\}_{n+1}^{FEM-BEM}=-\sigma^{BEM}\left\{\frac{\partial\varphi}{\partial n}\right\}_{n+1}^{BEM-FEM}\\ \left\{\frac{\partial\varphi}{\partial n}\right\}_{n+1}^{FEM-BEM}=-\frac{\sigma^{BEM}}{\sigma^{FEM}}\left\{\frac{\partial\varphi}{\partial n}\right\}_{n+1}^{BEM-FEM}\end{gathered} \tag{14}$$

As a result the Neumann boundary conditions are found that are applicable in the direction of normal vector on the FEM side of the BEM/FEM boundary.

3.4.3 Calculation of electric field in the FEM domain

In this step the electric field in the FEM domain is calculated. The Dirichlet or Neumann boundary conditions are specified for the FEM domain boundary except for the FEM-BEM boundary itself. Therefore, for the matrix system (2) in matrices $[H]^{FEM}$ the FEM-FEM contributions from "pure" FEM boundary and the FEM/BEM contribution from the mutual FEM/BEM boundary can be written in the following form:

$$[H]^{FEM}\cdot\begin{Bmatrix}\{\varphi\}_{n+1}^{FEM-FEM}\\ \{\varphi\}_{n+1}^{FEM-BEM}\end{Bmatrix}=\begin{Bmatrix}\{Q\}_{n+1}^{FEM-FEM}\\ \{Q\}_{n+1}^{FEM-BEM}\end{Bmatrix} \tag{15}$$

where matrix elements $\{Q\}_{n+1}^{FEM}$ are calculated using expressions (4) and (14). The outcome are potentials $\{\varphi\}_{n}^{FEM-BEM}$ at the FEM/BEM boundary.

3.4.4 Correction of calculated potentials at the BEM/FEM boundary

In this step correction of the potentials calculated in the previous step at the FEM/BEM boundary is done. The correction is made by using the successive underrelaxation method:

$$\{\varphi\}_{n+1}^{BEM-FEM} = (1-\theta)\cdot\{\varphi\}_{n}^{BEM-FEM} + \theta\cdot\{\varphi\}_{n+1}^{FEM-BEM} \tag{16}$$

Values of underelaxation θ are in range 0 – 1.

3.4.5 Check if the convergence criterion is fulfilled

4 Calculation results

The current density distribution over the structure being protected is given in figure 3. The potential distribution on the ground surface is shown in figure 4.

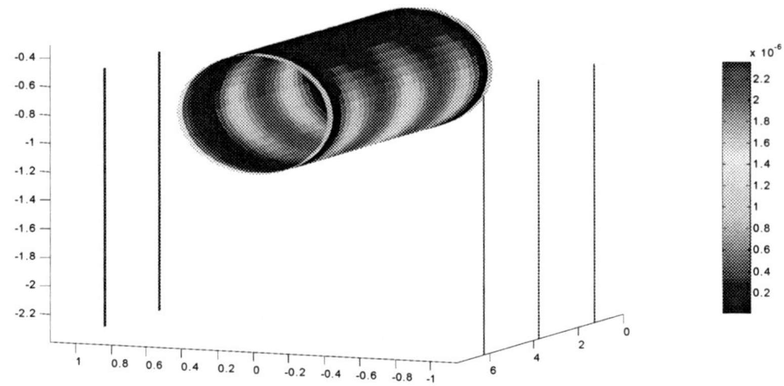

Figure 3: Current density distribution.

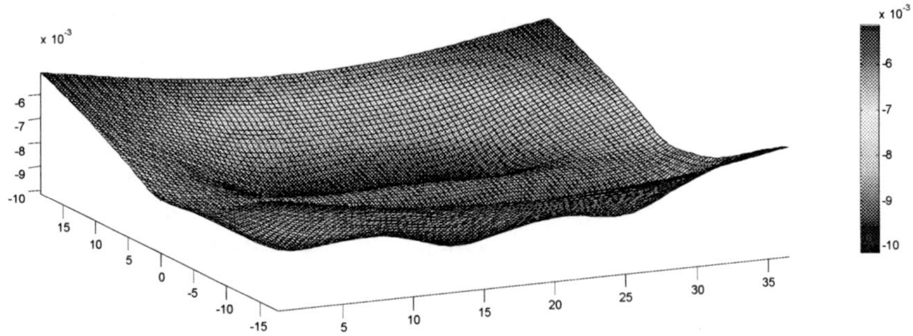

Figure 4: Potential distribution on the ground surface.

5 Conclusion

Application of the coupled BEM/FEM method has proved to be an efficient method for solving the problems in field of cathodic protection. Comparison numerical results have shown good agreement with the readings, protective potentials and current, taken on site.

Also presented algorithm has very stable convergence, and saves on computational time and memory requirements as well. Nonlinear boundary conditions are used in this software. The selection of boundary conditions is very delicate task, and choice of the parameters to be used is subject of serious analysis for specialists in field of electrochemistry.

References

[1] A. Muharemović, Distribution of the protection current and potential in system of cathodic protection with sacrificial anodes, The 5th Libyancorr. 2005.

[2] A. Muharemović, A contribution to the analysis of the polarization effects in optimization and dimensioning of a cathodic protection system, Sarajevo 1989.

[3] H. Zildžo, The numerical methods in power engineering, ETF Sarajevo 2004.

[4] Haznadar Z., Štih Ž., Electromagnetic fields, waves and numerical methods, Zagreb, 1997.

[5] D. S. Burnett, Finite element analysis from concepts to applications,Addison-Wesley Publishing Company, Massachusetts, 1987.

[6] Zienkiewicz O. C., The Finite Element Method, New York, McGraw-Hill, 1977.

[7] R.A. Adey, (ed). *Modelling of Cathodic Protection Systems*, Advances in Boundary Elements, Vol. 12, Southampton, Boston, WIT Press 2006.

[8] Kennelley, K.J., Bone, L. & Orazem, M.E., Current and potential distribution on a coated pipeline with holidays: Part 1. model and experimental verification. *Corrosion,* 49 (3), pp. 199–210, 1993.

[9] Nisançiöglu, K., Predicting the time dependence of polarization on cathodically protected steel in seawater, *Corrosion* , 43, pp 100–111, 1987.

[10] Elleithy, W. M., Al-Gahtani, H. J. And El-Gebeily, M., "Iterative Coupling of BE Methods in Elastostatics", Engineering analysis with Boundary Elements, Vol. 25. No. 8, August 2001, pp. 685–695.

[11] H.Zildžo and H. Matoruga, "Calculation of GIS 110 kV Insulating Bushing Apllying Hybrid BEM-FEM Method", XLII. International Scientific Conference on Information, Communication and Energy Systems Technologies – ICES 2007, Ohrid.

[12] H.Zildžo and H. Matoruga, "Calculation of transfered potentials in Substation applied Boundary element methods", 8. Konferenca Slovenskih elektroenergetikov- Sloko CIGRE – Čatež 2007.

Boundary element model of the human head exposed to an electrostatic field generated by Video Display Units

D. Čavka[1], D. Poljak[1], A. Peratta[2] & C. Brebbia[2]
[1]*University of Split, Croatia*
[2]*Wessex Institute of Technology, UK*

Abstract

The goal of this paper is to investigate the interaction between the electrostatic field of a Video Display Unit (VDU) and the human head located in front of it. Special attention is given to the field distribution over the surface of the face. The boundary element model for assessment of the electrostatic field, by solving Laplace equation for the electric potential, is implemented. The electrostatic field is calculated for two different faces and then compared to the results obtained by the other numerical methods. The advantage of the presented numerical method is a high resolution and a higher accuracy. The methodology development is equally applicable to other industrial electrostatic field problems.
Keywords: boundary element method, electrostatic field, human exposure, video display units.

1 Introduction

Video display units (VDUs) based on the cathode ray tube (CRT) are sources of several types of radiation e.g. X ray radiation, optical radiation (ultraviolet radiation, visible light and infrared radiation) electromagnetic radiation and electrostatic field. Along with the expanding use of VDUs some concerns about the effect of these fields on the human health have appeared. Over the years of work it has been noticed and proven that X ray radiation, optical radiation, high (~MHz) and low (~kHz) frequency electromagnetic fields are well below technical guidelines [1, 2] hence, considered not harmful for the health.

On the other hand, extremely low frequency (ELF) (~Hz) electromagnetic and electrostatic fields might be associated with some skin diseases, suppression of melatonin, or induction of phosphenes in the eyes, despite the fact that there is no

WIT Transactions on Modelling and Simulation, Vol 47, © 2008 WIT Press
www.witpress.com, ISSN 1743-355X (on-line)
doi:10.2495/BE080121

strong evidence of adverse health effects from domestic levels of ELF electromagnetic fields [1–3].

The present work deals with electrostatic fields, as large electrostatic fields have been detected from several types of VDUs and possibly related with skin rashes [3, 4]. Moreover, regarding health effects, the electrostatic fields are still neither fully investigated nor completely understood. To the best of authors' knowledge, no standards or guidelines exist for electrostatic fields. However, it has to be pointed out that the main concern due to exposure to these fields is linked with particle transport and deposition [5].

In the present paper, the realistic, three-dimensional, anatomically based model of the human head exposed to electrostatic field from VDU is presented. The electrostatic field around human head is assessed using the Boundary Element Method (BEM) [6, 7]. Special attention is given to the shape of the face by investigating two different faces. Furthermore, the field strength on the nose tip and eyes has been studied. Also, results obtained within this work are compared to those found by different methods, i.e. Finite Differences (FDM) [3] and Finite Elements (FEM) [8]. The principal motivation for studying this problem is lack of information of electrostatic field at the surface of human face, and the need to investigate accuracy of the results already obtained and published. Such information then can be used for solving problems like particle deposition on human face and eyes.

Another aim of the present work is to demonstrate that BEM is ideally suited to modelling of complex surfaces such as the human face, and that the results obtained are highly accurate.

2 Formulation of the problem

Assuming the charge density to be negligible, the mathematical description of the 3D electrostatic field between a VDU and human head is given by Laplace equation for electric potential φ:

$$\nabla^2 \varphi = 0 \tag{1}$$

with the associated boundary conditions:

$$\varphi = \varphi_s \text{ on the display,} \tag{2}$$

$$\varphi = \varphi_h \text{ on the head,} \tag{3}$$

$$\nabla \varphi \cdot \vec{n} = 0 \text{ on the far field boundaries.} \tag{4}$$

As shown in figure 1, the Dirichlet boundary conditions (2) and (3) are specified on the face and the display, while Neumann conditions (4) are imposed on all remaining exterior boundaries.

It is worth noting that the head is considered to be the perfect conductor thus being itself an equipotential surface with potential φ_h.

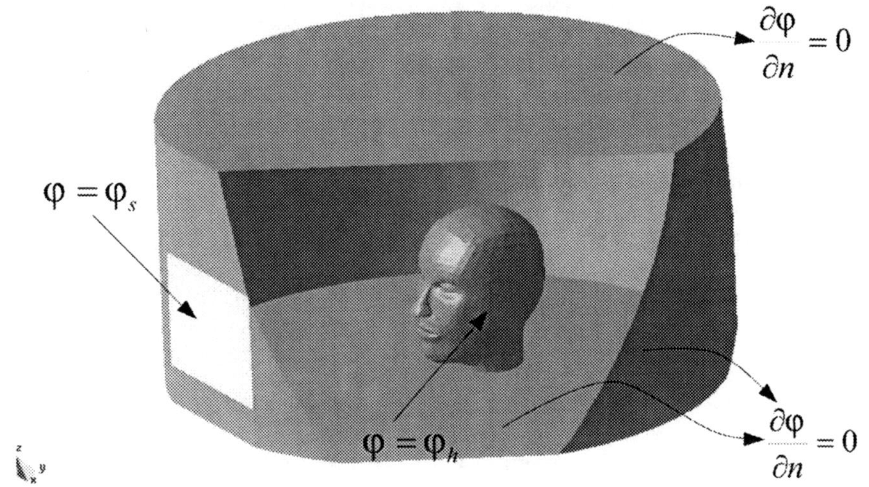

Figure 1: Geometry and boundary conditions for numerical 3D model of human seated in front of a VDU.

3 Model parameters and numerical solution

3.1 Model properties

Presented formulation involves the parameters l_s, d_s, φ_s and φ_h. They represent distance between display and nose tip, size of display (diagonal display size given in inches), electrostatic potential on display and electrostatic potential on head, respectively. The mean electric potential on a CRT monitors is in the range of 1–15kV. In this case the electric potential of the display is assumed to be relatively very high (15 kV) and it is considered as approximately worst-case scenario. In the present work standard conditions for the person in front of display are defined in table 1.

Table 1: Standard conditions.

Magnitude	l_s (m)	d_s (in)	φ_s (kV)	φ_h (kV)
Value	0.4	17	15	0

It is worth emphasizing that, the eyebrows are considered to have the same potential as the face. Other parameters such as temperature, humidity and conductivity of the surface of the screen glass are not considered in order to simplify the model. But, it has to be underlined that air humidity is very important factor, and in this calculations is considered very low (dry air).

Attention is given to shape of the face by investigating two different face geometries: person 1 (Figure 2 a)) and person 2 (Figure 2 b)).

The geometrical information of the models was introduced by means of the software 3D Studio Max, while the computational domain was preprocessed in GiD [9]. GiD is commercial software for professional pre and post processing developed by International Center for Numerical Method in Engineering, Barcelona, Spain. The postprocessing was performed using GiD, as well.

3.2 Boundary element solution

Applying the weighted residual approach, equation (1) can be integrated over the calculation domain Ω:

$$\int_{\Omega} \nabla^2 \varphi \, \psi \, d\Omega = 0 \tag{5}$$

where ψ is the weighting function. Using the following identities:

$$\nabla \cdot (\nabla \varphi \psi) = \psi \nabla^2 \varphi + \nabla \varphi \cdot \nabla \psi \tag{6}$$

$$\nabla \cdot (\varphi \nabla \psi) = \nabla \varphi \cdot \nabla \psi + \varphi \nabla^2 \psi \tag{7}$$

and applying the generalised Gauss theorem equation (5) can be written, as follows:

$$\int_{\Omega} \psi \nabla^2 \varphi d\Omega = \int_{\Gamma} \psi \frac{\partial \varphi}{\partial n} d\Gamma - \int_{\Gamma} \varphi \frac{\partial \psi}{\partial n} d\Gamma + \int_{\Omega} \varphi \nabla^2 \psi d\Omega \tag{8}$$

The weighting function ψ can be chosen to be the solution of the differential equation, i.e.

$$\nabla^2 \psi - \delta(\vec{r} - \vec{r}\,') = 0 \tag{9}$$

where δ is the Dirac delta function, r denotes the observation points and r' denotes the source points. The solution of equation (9) is called the fundamental solution or Green function.

Thus, the domain integral in (8) can be written:

$$\int_{\Omega} \varphi \nabla^2 \psi d\Omega = -\int_{\Omega} \varphi \delta(\vec{r} - \vec{r}\,') d\Omega = -\varphi_i \tag{10}$$

for any point inside the domain.

Combining equations (8)–(10) the following integral relation is obtained:

$$\varphi_i = \int_{\Gamma} \psi \frac{\partial \varphi}{\partial n} d\Gamma - \int_{\Gamma} \varphi \frac{\partial \psi}{\partial n} d\Gamma \tag{11}$$

The integral equation (11) can be regarded as the Green representation of the function φ.

The function ψ is the fundamental solution of the equation (9). For three-dimensional problems it is given by the expression:

$$\psi = \frac{1}{4\pi R} \tag{12}$$

where $R = |\vec{r} - \vec{r}'|$ is the distance from the source point (boundary point) to the observation point.

When the observation point i is located on the boundary Γ, the boundary integral becomes singular as R approaches zero. In order to deal with this singularity a small sphere for 3D problems is considered.

Calculating the surface integrals over the small spheres around singular points, finally follows the integral formulation of Laplace equation (1):

$$c_i\varphi_i = \int_\Gamma \psi \frac{\partial \varphi}{\partial n} d\Gamma - \int_\Gamma \varphi \frac{\partial \psi}{\partial n} d\Gamma \tag{13}$$

where

$$c_i = \begin{cases} 1, & i \in \Omega \\ 1/2, & i \in \Gamma \text{ (smooth boundary)} \end{cases} \tag{14}$$

Having computed the scalar potential distribution the electrostatic field can be determined from expression:

$$\vec{E} = -\nabla \varphi \tag{15}$$

3.3 Boundary discretisation

The three terms in equation (13) involve only boundary integrals. The boundary Γ can be discretised into N_e boundary elements, so that

$$\Gamma = \sum_{j=1}^{N_e} \Gamma_j \tag{16}$$

$$c_i\varphi_i = \sum_{j=1}^{N_e} \int_{\Gamma_j} \psi \frac{\partial \varphi}{\partial n} d\Gamma - \sum_{j=1}^{N_e} \int_{\Gamma_j} \varphi \frac{\partial \psi}{\partial n} d\Gamma \tag{17}$$

Each boundary element j contains a number (N_j) of collocation nodes, in which the potential or fluxes are evaluated. Thus, the values of the potential or its normal derivative at any point defined by the local coordinates (ξ_1, ξ_2) on a given boundary element can be defined in terms of their values at the collocation nodes, and the N_j interpolation functions Φ_k with $k = 1,..., N_j$ as follows:

$$\varphi(\xi) = \sum_{k=1}^{N_j} \Phi_k(\xi)\varphi_k \tag{18}$$

$$\frac{\partial \varphi(\xi)}{\partial n} = \sum_{k=1}^{N_j} \Phi_k(\xi) \frac{\partial \varphi_k}{\partial n} \tag{19}$$

Discretising the boundary and applying the collocation technique, expression (8) can be rewritten in the following form:

$$c_i\varphi_i = \sum_{j=1}^{N_e}\sum_{k=1}^{N_j}\left(\int_{\Gamma_j}\psi\,\Phi_k d\Gamma_j\right)\frac{\partial\varphi_{jk}}{\partial n} - \sum_{j=1}^{N_e}\sum_{k=1}^{N_j}\left(\int_{\Gamma_j}\frac{\partial\psi}{\partial n}\Phi_k d\Gamma_j\right)\varphi_{jk} \tag{20}$$

The notation in (20) can be simplified by making the use of matrix notation:

$$H[\varphi] = G\left[\frac{\partial\varphi}{\partial n}\right] \tag{21}$$

where

$$H_{il} = \delta_{il}c_i + \int_{\Gamma_j}\frac{\partial\psi_i}{\partial n_j}\Phi_k d\Gamma_j \tag{22}$$

$$G_{il} = \int_{\Gamma_j}\psi_i\,\Phi_k\,d\Gamma_j \tag{23}$$

where index l, identifying a collocation node within the domain, can be calculated in terms of indices j and k, by means of the nodal connectivity of the mesh, i.e. index $l = 1,\ldots,M$. and M, is the total number of collocation nodes in the domain. Basically, the index l is used to identify one of the adjacent freedom (collocation) nodes from a global point of view, and is given as a function of the indicator of element (j), and the local collocation node of that element (k). In the case of discontinuous collocation nodes:

$$M = \sum_{j=1}^{N_e} N_j \tag{24}$$

The boundary differential $d\Gamma_j$ can be expressed in terms of the domain local coordinates (ξ) via the jacobian of the transformation $|J|$:

$$d\Gamma_j = |J|\,d\xi_1\,d\xi_2 \tag{25}$$

Finally, equation (21) can be written as a given vector in function of the source term, or a characteristic matrix in function of the unknown potentials and normal fluxes at the collocation nodes of the boundary, the application of the prescribed boundary conditions and the assembly of the linear set of equations, that (25) produces, yields to a determined system of equations of dimensions N by N of the form:

$$\mathbf{A}\,\mathbf{x} = \mathbf{b} \tag{26}$$

where the vector of unknowns **x** contains the potentials and normal fluxes that were not prescribed as boundary conditions, the matrix A involves the coefficients of H and G, and the right hand side term involves the source term, and the boundary conditions.

4 Results and discussion

The mesh generated for the model of person 1 contains around 4500 constant elements, while for the model of person 2 around 6000 (figure 2). The difference is due to number of surfaces that form particular head. Numerical calculation was carried out using MATLAB code.

Figure 3 shows the electrostatic field E [V/cm] on the face of person 1 and person 2.

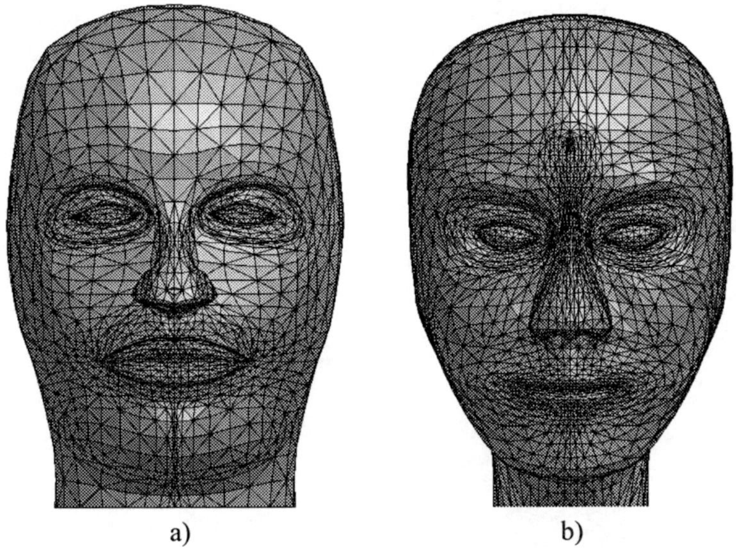

Figure 2: BEM discretisation of the heads: a) person 1; b) person 2.

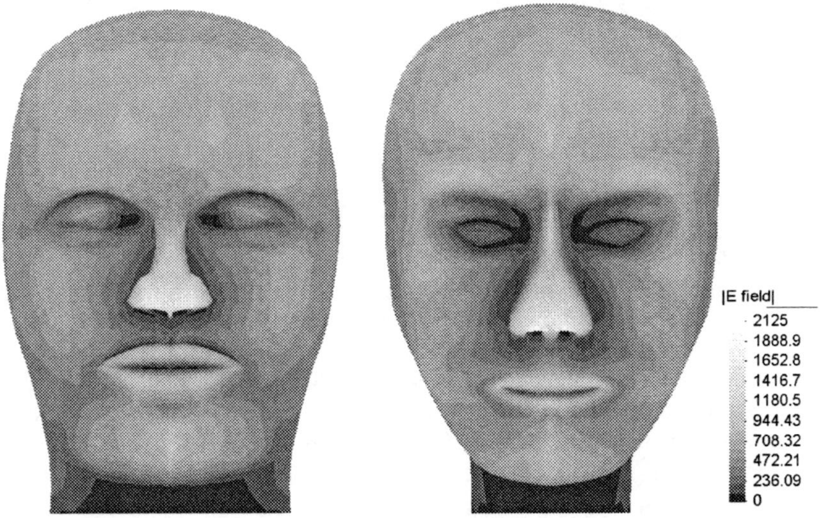

Figure 3: Electrostatic field strength on the face of persons.

The results show very different values of field strength in the two cases, i.e. the maximal value of the field at the nose of the person 1 (1760 V/cm=176 kV/m) is lower then the value for the person 2 (2125 V/cm). The difference is caused by the different shape of the face, particularly the shape of the nose. For most of the face area, the difference is not so significant. Certain difference is noticed on the forehead and brows, around 300 V/cm. Relatively high values, compared to the rest of the face, of electrostatic field can be found on the lips, chin and on the top of the cheek, around 1000 V/cm.

These details and differences are clearly visible from the perspective view of faces around nose and eyes, as shown in figure 4.

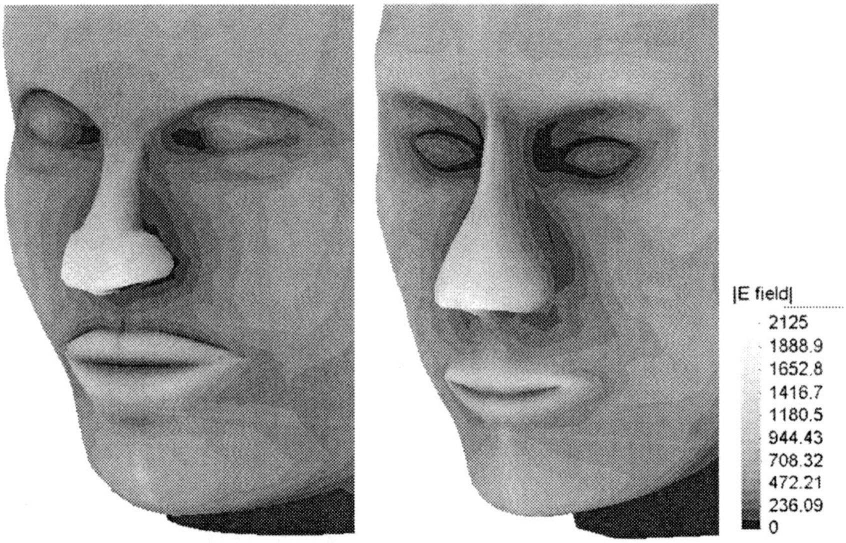

Figure 4: Electrostatic field strength [V/cm] around the nose and eyes.

It can be observed from figure 4 that the field strength is highly dependent on the shape of the face especially of the nose. Also from the same figure is clear that field strength around eyes is relatively low comparing to the rest of the face, around 400-500 V/cm.

Results obtained in this work are comparable with the ones obtained by different methods and presented in earlier publications such as [3] (FDM) and [8] (FEM). Exact comparison with results obtained in [3] cannot be made due to different shapes of the heads being considered. What should be stated is that the values of electrostatic field are very similar and the behaviour of the field on the face is similar in nature. Comparing results obtained in this work with the results obtained in [8] using FEM for the same head models some differences arise.

On the FEM model some "hot spots" (i.e. regions of high electrostatic field strength) can be noticed on the nose, having much higher values of the field than the same spots on the BEM model. Although, there are other parts where the values of the BEM are higher than those of the FEM.

Nevertheless, overall both methods give very similar results. The differences can be explained because of the difficulty of defining sufficiently refined FEM meshes around complex geometries such as the human face.

5 Conclusions

This paper investigates the interaction between the electrostatic field generated by VDUs and the human head.

As different from previous studies using Finite Differences (FDM) or Finite Elements (FEM), this work applies the Boundary Element Method (BEM), which generally is more accurate and versatile than the other two, allowing for a better representation of the shape of the human face.

Two different facial shapes were investigated and results compared with those obtained by FEM. Although both results were of similar accuracy the BEM allows for a better description of the human face.

Future work will include other factors such as air humidity and more refined models for area of complex geometry such as the ears. The mesh will also be refined in the region around the eyes introducing eyelashes and eyebrows. The work presented in this paper is important not only to study the electrostatic field effects of VDU but also of a whole range of other industrial equipments.

References

[1] International Commission on Non-Ionizing Radiation Protection (ICNIRP) and International labour organization: "Visual display units: Radiation protection guidelines", International labour office, Geneva 1994.

[2] Institute of Electrical and Electronics Engineers, *Biological and health effects of electric and magnetic fields from video display terminals,* IEEE Engineering in Medicine and Biology Magazine 16(3): 87–92, 1997.

[3] Nielsen, N.F., Michelsen, J., Michalsen, J.A.. & T. Schneider, Numerical calculation of electrostatic field surrounding a human head in visual display environments", *Journal of electrostatic,* **36**, pp. 209–223, 1996.

[4] Australian Radiation Protection and Nuclear Safety Agency, *Radiation Emission from Video Display Terminals*, Australian Radiation Protection and Nuclear Safety Agency, 2003.

[5] Nielsen, N.F. & Schneider, T., Particle Deposition onto a Human Head: Influence of Electrostatic and Wind Fields, *Bioelectromagnetics* **19**, pp. 246–258, 1998.

[6] Brebbia, C.A. & Dominguez, J., *Boundary Elements, An Introductory Course,* second edition. WIT Press: Boston, Southampton 1998.

[7] Gonzalez, C., Peratta, A. & Poljak, D., Boundary Element Modelling of the Realistic Human Body Exposed to Extremely-Low-frequency (ELF) electric Fields: Computational and Geometrical Aspects, *IEEE Transactions on Electromagnetic Compatibility,* Vol. **49**, No **1** pp. 153–162, February 2007.

[8] Čavka, D., Poljak, D. & Peratta, A., *Finite element model of the human head exposed to electrostatic field generated by Video Display Units*, SoftCOM conference proceedings, Split 2006.
[9] GiD 7 Reference Manual. International Center for Numerical Methods in Engineering. CIMNE. Spain. 2006. (http://gid.cimne.upc.es/).

Section 4
Damage mechanics and fracture

Interfacial debonding and damage progression in particle-reinforced composites

S. Womack & M. S. Ingber
Department of Mechanical Engineering, University of New Mexico, Albuquerque, NM, USA

Abstract

Decohesion between the matrix and inclusions in advanced multi-component composite materials is an important failure mechanism. In particular, the overall bulk strength of the composite depends strongly on interfacial damage and debonding. A 3-D model of interfacial decohesion is developed for a spherical particulate-reinforced composite material. In this model, interfacial failure is initiated by the loss of ellipticity. A discrete constitutive equation is used to represent failure at the interface of the composite. Incremental displacement steps are applied to the model within the elastic range of the constituent materials and the interface is analyzed using the boundary element method to determine the progression of interfacial damage and debonding. The effective properties of the bulk composite are calculated at each incremental step.
Keywords:composite materials, decohesion model, incremental debonding, boundary element method.

1 Introduction

A composite is a manufactured material with two or more physically or chemically distinct phases. Particle composites typically contain a matrix material and inclusions. The interface between the inclusion and the matrix is a bonding surface, across which both weak and strong discontinuities occur [1]. Weak discontinuities are defined as discontinuous changes in the strain field across the limits of a narrow strain localization region over which the displacement field remains continuous. Under load, strong discontinuities may develop, causing the displacement field to become discontinuous and the strain field to become unbounded.

WIT Transactions on Modelling and Simulation, Vol 47, © 2008 WIT Press
www.witpress.com, ISSN 1743-355X (on-line)
doi:10.2495/BE080131

Composites are designed to create desirable material properties such as increased strength, increased stiffness, decreased weight, corrosion resistance, and increased or decreased thermal and electrical conductivity. The overall performance of a composite depends on the material properties of each phase as well as the interfacial properties between the matrix and inclusions [2]. These material and interfacial properties govern how the material fails, including such failure modes as brittle fracture, ductile rupture, delamination, fiber pullout, yielding, and excessive deformation [3]. Composites commonly fail along the interface between the matrix and inclusion. This type of failure is called interfacial material failure and is defined as the formation of two new surfaces from a previously bonded interface between two materials. Interfacial decohesion is usually observed in composites with very low strength matrices relative to the inclusion, (i.e., ceramic inclusions in a pure aluminum matrix), while particle fracture usually occurs with a medium to high strength matrix [4, 5].

Given the significance of interfacial damage progression on the bulk strength and toughness of composite materials, it is not surprising that one of the major research areas in composites is the modeling of bonding interfaces between phases. The interfacial zone has been modeled in a number of ways, including as a narrow region of continuum with graded properties, as an infinitely thin surface with springs, and as a cohesive zone with traction-separation relations. Recently, the cohesive zone approach has become widely used. Needleman [6] was one of the first to apply the traction-separation cohesive relationship to the model. Needleman focused on a model that describes the evolution from initial debonding to decohesion and subsequent void growth. Needleman suggested a continuum approach where strain is defined not only in the continuous part of the body, but also at the discontinuous interface, and therefore, standard stress-strain constitutive equations can be considered everywhere. The major area of research using this technique is the determination of the traction-separation cohesive relationship, which has been modeled using various functions including linear [7], quadratic [8], polynomial [6], trapezoidal [9], bilinear [10], and exponential functions [11, 12].

Schreyer *et al.* [3] took a fundamentally diferent approach than Needleman, in which strain softening occurs after ellipticity is lost. In this approach, the onset of inelasticity or the loss of strong ellipticity are indicators of the progression of more complex material behaviors that lead to the progressive softening of the material [13]. The direct calculation of the strong discontinuity in displaceent can be determined because the discrete equation can be applied at the instant ellipticity is lost. Once decohesion is initiated on a surface of discontinuity, the adjacent continuum tends to unload into the elastic regime.

Han *et al.* [14] combined Schreyer's decohesion model with the boundary element method to develop a 2D simulation capability of interfacial decohesion of a fiber-reinforced composite material. The current research extends the work of Han *et al.* from 2D fiber-reinforced composites to 3D reinforced composites. In particular, boundary element analysis will be used in conjunction with the damage function proposed by Schreyer [3] at the interface between the matrix and inclusion to study interfacial damage of the composite.

2 Numerical formulation

The computational domain considered in the current research is comprised of a matrix material with an embedded spherical inclusion. A boundary integral equation (BIE) is written for each of the two zoned-homogeneous, isotropic, elastic regions. Collocating the BIE at the boundary element nodes results in a system of linear equations relating the components of traction to the components of displacement. These equations can be represented symbolically as

$$[H^k]\{\mathbf{u}^k\} = [G^k]\{\mathbf{t}^k\} \tag{1}$$

where the superscript k represents the region number, and $\{\mathbf{u}^k\}$ and $\{\mathbf{t}^k\}$ represent the components of displacement and traction, respectively, at the collocation nodes.

On the outer boundary of the matrix material, either the displacement or the traction is prescribed. On the interfacial nodes between the spherical inclusion and matrix, neither the displacement nor traction components are known *a priori*. The system of equations is closed by setting the following interfacial boundary conditions.

$$\{u^i_j\} - \{u^m_j\} = [u] \tag{2}$$

$$\{t^i_j\} = -\{t^m_j\} \tag{3}$$

where $[u]$ represents a jump in displacement along the interface as discussed in the next section, the superscript i represents the inclusion and superscript m represent the matrix, and the subscript represents the global node within the appropriate region.

Initially, the interface between the matrix material and the inclusion is assumed to be perfectly bonded, that is, continuity of traction and displacement is assumed along the interface. In order to minimize the computation cost of the analysis, the displacement at which damage initiates is calculated and the decohesion analysis is started at that displacement. The displacement is then incrementally increased to study the progression of damage.

At each increment of displacement, the boundary integral equations are solved and the tractions along all interfaces are evaluated. The development of damage (decohesion) is determined through the use of a damage function F given by

$$F = \left[\left(\frac{\tau_n}{\tau_{nf}}\right)^2 + \left(\frac{\tau_t}{\tau_{tf}}\right)^2\right]^{\frac{1}{2}} - f \quad \text{when} \quad \tau_n \geq 0$$

$$F = \left|\frac{\tau_t}{\tau_{tf}}\right| - f \quad \text{when} \quad \tau_n < 0 \tag{4}$$

where τ_n is the normal component of traction, τ_t is the tangential component of traction, τ_{nf} is the value of failure initiation traction in a pure tensile mode, τ_{tf} is the value of failure initiation traction in a pure shear mode, and f is the so-called softening function. The function F is defined so that no damage occurs if

$F < 0$ and $F > 0$ is not allowed. Damage develops only if $F = 0$. The softening function f is chosen so that $f = 1$ for an undamaged interface and $f = 0$ for a fully decohered interface. For $0 < f < 1$, a linear relationship is chosen between the absolute value of the jump in displacement $[\mathbf{u}]$ at the interface so f is given by

$$f = 1 - \frac{|[\mathbf{u}]|}{u_0} \tag{5}$$

where the model parameter u_0 is the value of $|[\mathbf{u}]|$ at which complete decohesion has occurred. Although the choice of f is somewhat arbitrary, in the special case of uniaxial tension, the softening function translates into a curve of normal traction versus normal displacement discontinuity in which case the area under the curve is the fracture energy, G_f.

After each increment of displacement, the damage function F is evaluated at each interfacial node. If $F > 0$, the threshold to initiate decohesion has been exceeded and is not allowed by the decohesion model. As an example, if the traction at an interfacial node had no tangential component, $F > 0$ would imply that τ_{nf} had been exceeded. In order to drive F to within a specified tolerance ϵ of 0, the following algorithm is used. Nodes with the largest positive values of the damage function are chosen for softening, i.e., given a jump in displacement across the node between the inclusion and matrix.

The displacement discontinuities at the softened nodes are determined by the following evolution equation

$$\Delta[\mathbf{u}] = \Delta\lambda\mathbf{g} \tag{6}$$

where $[\mathbf{u}] = \{[\mathbf{u}]_n^2 + [\mathbf{u}]_t^2\}^{\frac{1}{2}}$, $\mathbf{g} = \frac{\partial F}{\partial \tau}$, and $\Delta\lambda = \delta\lambda_1 + \delta\lambda_2 + \cdots$. The secant algorithm is applied to determine $\Delta\lambda$ in order to bring F to within a tolerance ϵ of 0. An initial small value for $\delta\lambda_1$ is assumed. This then provides the initial assumed value for $\Delta[\mathbf{u}]$ as shown in Eq. 6. This jump in displacement provides stress relief which is quantified by performing the boundary element analysis with the modified boundary conditions (Eq. 2). The damage function is then re-evaluated at the softened nodes. If F is still positive and above the tolerance, $\delta\lambda_2$ is determined using the equation

$$\delta\lambda_2 = \delta\lambda_1 \frac{F_1}{F_0 - F_1} \tag{7}$$

where the subscript on F indicates the iteration number. The procedure is continued until $F < \epsilon$. During any iteration, if $|[\mathbf{u}]|$ reaches u_0, the boundary condition at that node is set to traction free indicating that the node is completely decohered.

3 Damage and debonding propagation

For all material tests, the outside dimensions of the material specimen is a cube with edge length 2 centered at the origin. The outside matrix material is aluminum with one spherical ceramic boron particle at a volume fraction of 10%. The elastic material properties are given by $E_{al} = 67.51$ GPa, $\nu_{al} = 0.35557$, $E_B = 413.04$ GPa, and $\nu_B = 0.2$.

Two types of material tests are performed in this research, namely, a tensile test and a shear test. For the tensile test, an incremental uniform tensile displacement is applied to the top and bottom surfaces of the composite specimen. The effective Young's modulus for the composite undergoing the tensile test is calculated as

$$E_{eff} = \frac{F_n L}{A\delta_n} \tag{8}$$

where F_n is the resultant normal force on the top or bottom surface of the specimen, L is the specimen length, A is the cross-sectional area, and δ_n is the applied normal displacement. The resultant force is calculated by integrating the tractions over the top and bottom faces.

For the shear test, an incremental uniform shear displacement is applied to the top and bottom surfaces of the composite specimen. The effective shear modulus is calculated as

$$G_{eff} = \frac{F_s L}{A\delta_s} \tag{9}$$

where, in this case, F_s is the resultant shear force on the top or bottom surface of the specimen and δ_s is the applied shear displacement.

The maximum interfacial traction is defined as

$$\tau_f = \sqrt{\tau_{nf}^2 + \tau_{tf}^2} \tag{10}$$

where τ_{nf} and τ_{tf} are the normal and shear mode failure limits used to calculate the damage function F. For all cases, the maximum interfacial traction is chosen as $\tau_f = 12$ MPa. The parameter u_o, the maximum value that the displacement discontinuity reaches before debonding is complete, is chosen as $u_o = 0.000046$, again for both the ductile and brittle failure models. The values of τ_f and u_o are chosen to ensure that the interfacial bond will substantially fail before the matrix material yields.

Two cases are considered for both the tensile and shear test, namely, $\tau_{nf}/\tau_{tf} = 0.5$ and $\tau_{nf}/\tau_{tf} = 5.0$ to illustrate the difference between normal-force dominated failure and shear-force dominated failure.

3.1 Tensile load results

The first case considered is the single inclusion model undergoing a tensile load. Upon application of a normal displacement, the traction distribution on the top and bottom surfaces of the composite is non-uniform because of the inclusion. The magnitude of the traction vector normalized by the maximum interfacial traction, τ_f, is plotted for three values of the applied strain in Fig. 1. As shown in Fig. 1(a) for the case of a composite consisting of an aluminum matrix with embedded ceramic spherical particle before the onset of interfacial damage, the traction distribution across the top surface increases towards the center of the surface

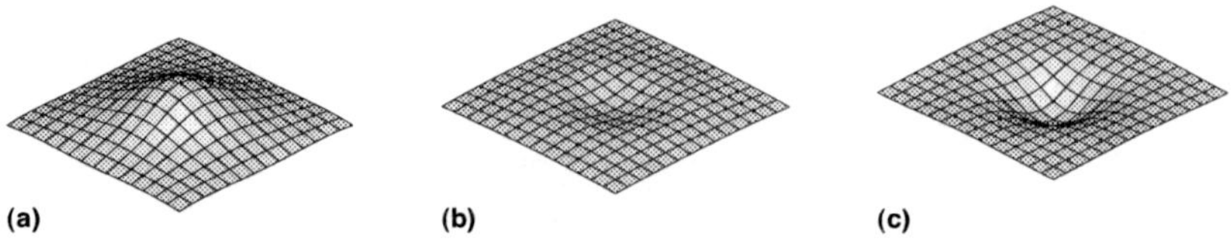

Figure 1: The magnitude of the traction vector along the top surface of the composite specimen undergoing a tensile test (a) before the initiation of interfacial damage, (b) after the initiation of interfacial damage but before the onset of debonding, and (c) after the onset of debonding.

where the gap between the surface and inclusion is smallest indicating the higher Young's modulus of the inclusion. The traction distribution shown in Fig. 1(b) for the case in which decohesion has been initiated but before any debonding has occurred is quite different. The damage along the interface at the north pole of the inclusion has reduced the load bearing capability resulting in a greatly reduced normal traction in that region. To compensate, the normal traction is increased in the adjacent nodes. Finally, as seen in Fig. 1(c) for the case in which debonding has occurred at the north pole, the normal traction in this region has been further reduced. Although the plots shown are for the normal-force dominated failure, the results for the shear-force dominated failure are qualitatively similar.

The normalized effective Young's modulus is defined by $E_n = E_{eff}/E_{mat}$ where E_{mat} is the Young's modulus of the matrix material. The normalized Young's modulus is shown as a function of the applied normal strain in Fig. 2 for the shear-force dominated and normal-force dominated cases undergoing a tensile load. Initially,

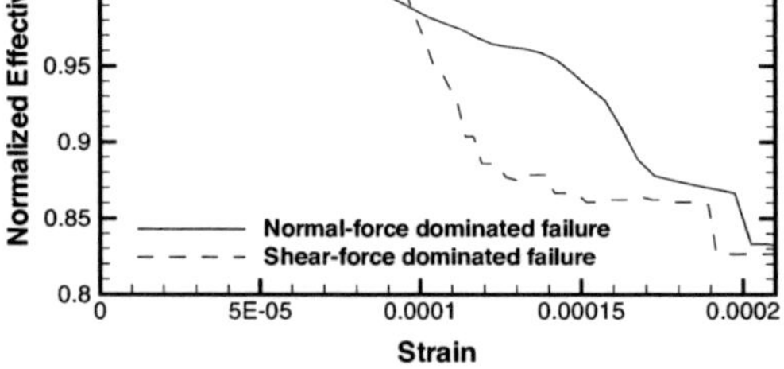

Figure 2: The normalized effective Young's modulus for shear- and normal-force dominated failure undergoing tensile loading.

the normalized Young's modulus is given by $E_n = 1.165$, indicating the enhanced stiffness of the composite caused by the spherical inclusion. As the incremental displacement is applied, the interface of the composite begins to experience damage when the bulk strain reaches 17 $\mu\epsilon$ ($\mu\epsilon = 10^{-6}$ mm/mm) and 64 $\mu\epsilon$ for normal-force and shear-force dominated failure, respectively. Debonding of the first node occurs at approximately 100 $\mu\epsilon$ regardless of the failure mode. The normalized Young's modulus plateaus after the stabilization of the equatorial region, at approximately 200 $\mu\epsilon$ for the normal-force dominated failure and 160 $\mu\epsilon$ for shear-force dominated failure.

There is a notable difference between the shear-force dominated and normal-force dominated failure modes. In the shear-dominated failure mode, damage initiates at either the north or south pole and then progresses from that pole towards the equatorial compressive zone before damage initiates at the opposite pole. The damage occurs only in one hemisphere from a strain of 64.0 $\mu\epsilon$ until approximately 122 $\mu\epsilon$. At that point, the effective Young's modulus remains relatively constant until a strain of approximately 142 $\mu\epsilon$ as seen in Fig. 2. Beyond this strain, damage is initiated at the pole of the other hemisphere, and now damage and subsequent debonding progresses within the new hemisphere. For normal-force dominated failure, the progression of failure occurs simultaneously in both hemispheres.

3.2 Shear load results

The second case considered is the single inclusion model undergoing a shear test. When a shear displacement is applied to the top surface of the composite the traction distribution on this surface is again non-uniform. The magnitude of the traction vector normalized by the maximum interfacial traction is plotted for a case in which damage has been initiated in Fig. 3. For the shear test, damage initiates at a location 45° from the vertical. The right-hand side of the top surface in the figure is the side closest to the damaged interface, and hence, the tractions are lower along the right side of the top surface compared to the left side because of the reduced load bearing capacity.

The normalized effective shear modulus is defined by $G_n = G_{eff}/G_{mat}$ where G_{mat} is the shear modulus of the matrix material. The normalized shear modulus is shown as a function of the applied shear strain in Fig. 4 for the shear-force dominated and normal-force dominated cases undergoing the shear test. Initially, $G_n \approx 1.10$, indicating an increased shear modulus of the composite compared to the matrix material. As incremental shear displacement is applied, the interface of the composite begins to undergo softening when the bulk strain reaches 53 $\mu\epsilon$ and 110 $\mu\epsilon$ for normal- and shear-force dominated failure, respectively. Debonding of the first node occurs at approximately 228 $\mu\epsilon$ and 200 $\mu\epsilon$ for normal- and shear-force dominated failure. The normalized shear modulus plateaus after the stabilization of the rotated equatorial region, at approximately 268 $\mu\epsilon$ for normal-dominated failure and 210 $\mu\epsilon$ for shear-dominated failure. As in the case of the tensile test, damage progresses in one hemisphere at a time for the shear-dominated

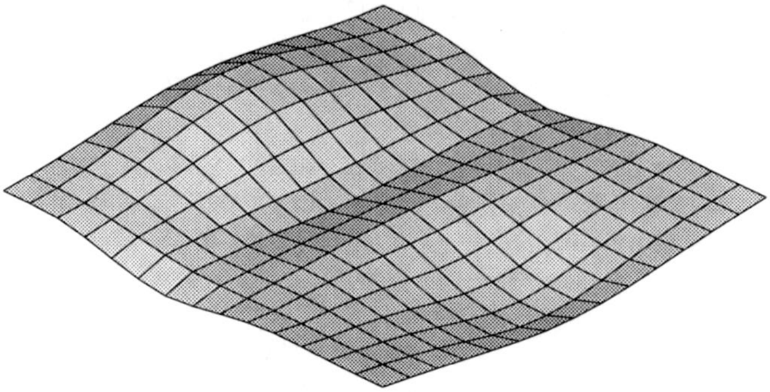

Figure 3: The magnitude of the traction vector along the top surface of the composite specimen undergoing a shear test after the onset of damage.

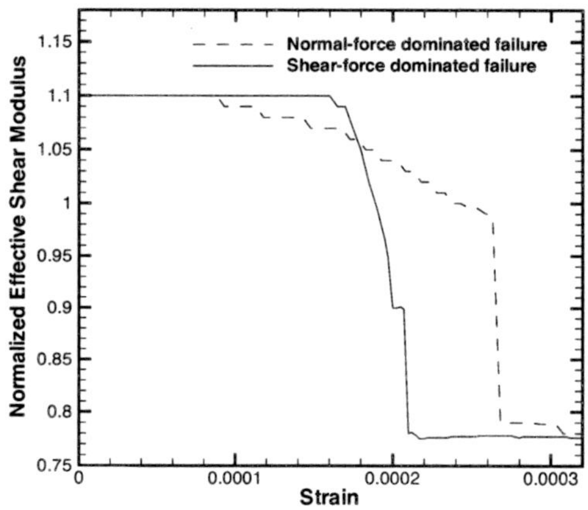

Figure 4: The normalized effective shear modulus for shear- and normal-force dominated failure undergoing shear loading.

failure case whereas damage progresses simultaneously in both hemispheres for the normal-dominate failure case.

4 Conclusions

A softening decohesion model has been combined with linear elasticity to describe the evolution of interfacial failure in spherical inclusion reinforced materials. As softening occurs, the sum of the work performed by the traction and the stored energy released by the material equals the energy dissipated. Progressive evolution of the debonding zones and the propagation of decohesion along the interface have been illustrated by incremental simulations. The material failure parameters, τ_{nf} and τ_{tf}, play an important role in the decohesion process and may explain the difference between failure in ductile and brittle materials. In particular, if the ratio

τ_{tf}/τ_{nf} is small, then the material is more likely to fail in shear. In these cases once a portion of the interface becomes partially decohered, the damage zone slides tangentially along the interface until a compressive region is reached. On the other hand, if τ_{tf}/τ_{nf} is large, then this shear mode failure is inhibited and damage tends to progress simultaneously on both sides of the inclusion. Decohesion has significant effect on the macroscopic response of the composite materials. In particular, the effective composite properties are adversely affected to the point where the partially decohered inclusion can actually diminish the effective shear and Young's moduli.

References

[1] K. K. Chawla, Composite Materials Science and Engineering, Springer Science+Business, New York, 1998.

[2] C. R. Dobert, R. Mahnken, E. Stein, Numerical simulation of interface debonding with a combined damage/friction constitutive model, Comp. Mech. 25 (2000) 456–467.

[3] H. L. Schreyer, D. L. Sulsky, S. J. Zhou, Modeling delamination as a strong discontinuity with the material point method, Comp. Meth. Appl. Mech. Engr. 191 (2002) 2483–2507.

[4] A. F. Whitehouse, T. W. Clyne, Cavity formation during tensile straining of particulate and short fiber metal-matrix composites, Acta Met. Mat. 41 (6) (1993) 1701–1711.

[5] L. Babout, E. Maire, R. Fougeres, Damage initiation in model metallic materials: X-ray tomography and modeling, Acta Mat. 52 (2004) 2475–2487.

[6] A. Needleman, A continuum model of void nucleation by inclusion debonding, J. Appl. Mech. 54 (1987) 525–531.

[7] G. T. Camacho, M. Ortiz, Computational modeling of impact damage in brittle materials, Int. J. Solids Struct. 33 (20-22) (1996) 2899–2938.

[8] V. Tvergaard, Effect of fiber debonding in a whisker-reinforced metal, Mat. Sci. Engr. 125 (2) (1990) 203–213.

[9] V. Tvergaard, J. W. Hutchinson, The relation between crack-growth resistance and fracture process parameters in elastic plastic solids, J. Mech. Phys. Solids 40 (6) (1992) 1377–1397.

[10] J. Segurado, J. LLorca, A new three-dimensional interface finite element to simulate facture in composites, Int. J. Solids Struct. 41 (2003) 2977–2993.

[11] A. Needleman, An Analysis of Decohesion Along an Imperfect Interface, Int. J. Fracture 42 (1) (1987) 21–40.

[12] A. Salvadori, A symmetric boundary integral formulation for cohesive interface problems, Comp. Mech. 32 (2003) 381–391.

[13] J. Oliver, Modeling strong discontinuities in solid mechanics via strain softening constitutive equations Part 1, Int. J. Num. Meth. Engr. 39 (1996) 3575–3600.

[14] R. Han, M. S. Ingber, H. L. Schreyer, Progression of failure in fiber-reinforced materials, Comp. Mats. Cont. 4 (3) (2006) 163–176.

Trefftz-type approach for 2D fracture problems

A. N. Galybin
Wessex Institute of Technology, Southampton, UK

Abstract

This study is aimed at the development of a Trefftz-type method for solving plane elastic boundary value problems for open contours, which models crack propagation in brittle materials. The idea of the approach is as follows. Complex potentials are sought as linear combinations of independent holomorphic functions with the weights addressing singularities at the crack tips. Then the collocation method is applied to satisfy boundary conditions, which leads to a linear system for determination of unknown coefficients in the representation for complex potentials. The system is, in general, overdetermined and, thus, the SVD regularisation is applied to find its approximate solution. Two examples are presented.

Keywords: cracks, complex potentials, Trefftz method, collocations, ill-posed problems.

1 Introduction

Common technique for solving plane elastic problems with cracks assumes determination of two holomorphic functions (complex potentials, see Muskhelishvili [1]) that, in general, have square root singularities at crack tips. Usually, complex potentials are found by solving singular integral equation, SIE, (or a system of SIEs) with respect to unknown generalised crack opening displacements or their densities. This approach requires discretisation of the boundary and application of special quadratures for singular integrals, which represent two common steps in boundary integral methods. The present study is aimed to avoid these two steps and to apply directly the Trefftz approach by representing complex potentials as linear combinations of known holomorphic functions.

WIT Transactions on Modelling and Simulation, Vol 47, © 2008 WIT Press
www.witpress.com, ISSN 1743-355X (on-line)
doi:10.2495/BE080141

The Trefftz approach for elastostatics is known and its different variants have been discussed [2- 6]. However application of the method for fracture mechanics problems (e.g., [4,5]) is not fully developed. Here the approach used in [6-7] for smooth contours and non-classical formulations is extended for the case of plane bodies with cracks. It is proposed to seek representations for complex potentials as linear combinations of independent holomorphic functions, that have square root singularities at the crack ends, followed by the determination of unknown (complex) coefficients from the boundary conditions by the collocation method. In general, the number of collocation points is assumed to be greater than the number of the sought coefficients, which results in an overdetermined system of linear algebraic equations. An approximate solution of the system is obtained by regularisation of the matrix by using the singular value decomposition method. The regularisation is important because it ensures that only linearly independent holomorphic functions are used as the basis functions in series for complex potentials. Two examples that use different sets of basis functions are presented to illustrate the approach. These are for a crack in a plane and acrack in a circle.

2 Trefftz approach for cracks in 2D elastic media

2.1 General expressions for the Trefftz approach using complex potentials

General solution (no body forces) for plane elastic domain has the form [1]

$$\frac{\sigma_{11}+\sigma_{22}}{2}=P(z,\bar{z})=\Phi(z)+\overline{\Phi(z)}$$

$$\frac{\sigma_{22}-\sigma_{11}}{2}+i\sigma_{12}=D(z,\bar{z})=\bar{z}\Phi'(z)+\Psi(z) \tag{1}$$

$$2G(u_1+iu_2)=W(z,\bar{z})=\kappa\varphi(z)-z\overline{\Phi(z)}-\overline{\psi(z)},\quad \Phi(z)=\varphi'(z), \Psi(z)=\psi'(z)$$

Here φ, ψ, Φ and Ψ are arbitrary holomorphic functions that are to be found from boundary conditions; harmonic function P and complex-valued function D represent mean stress and stress deviator which are functions of stress components σ_{ij}; W is a complex-valued function proportional to the displacement vector (u_1,u_2); G is the shear modulus, κ=3-4ν for plane strain and κ=(3-ν)/(1+ν) for plain stress, ν is Poisson's ratio.

Boundary value problem for cracks in plane elastic bodies is formulated as the first boundary problem of elasticity for which tractions (or stress vector) is known on the crack. In terms of stress functions and complex potentials this problem can be presented in the following form [1]

$$P(\zeta)+\frac{\overline{d\zeta}}{d\zeta}\overline{D(\zeta)}=2\operatorname{Re}(\Phi(\zeta))+\frac{\overline{d\zeta}}{d\zeta}\left(\zeta\overline{\Phi'(\zeta)}+\overline{\Psi(\zeta)}\right)=N(\zeta)+iT(\zeta),\quad \zeta\in\Gamma \tag{2}$$

Here $N(\zeta)$ and $T(\zeta)$ are normal and shear components of stresses on the boundary Γ of the considered domain Ω (interior or exterior including the crack); $P(\zeta)$ and $D(\zeta)$ are boundary values of the stress functions defined via boundary values of complex potentials $\Phi(\zeta)$ and $\Psi(\zeta)$. In fracture mechanics, in general, condition (2) is valid on the contour of the crack only, while on the rest of the boundary one can use mixed boundary conditions assuming that either tractions or displacements are known on different parts. Here, for the sake of simplicity, we apply (2) to the entire contour Γ.

Complex potentials $\Phi(z)$ and $\Psi(z)$ are sought in the form

$$\varphi(z)=\sum_{k=1}^{n} a_k h_k(z),\quad \psi(z)=\sum_{k=1}^{n} b_k h_k(z) \tag{3}$$

where $h_k(z)$ are independent functions holomorphic in the entire domain Ω; $2n$ complex coefficients a_k and b_k are unknown (for simplicity the number of terms in both sums are the same). To satisfy (2) one needs to find derivatives of (3); they can be presented in the form

$$\Phi(z)=\sum_{k=1}^{n} a_k H_k(z),\quad \Phi'(z)=\sum_{k=1}^{n} a_k H_k'(z),\quad \Psi(z)=\sum_{k=1}^{n} b_k H_k(z) \tag{4}$$

where $H_k(z)=h_k'(z)$ are holomorphic in Ω.

Substitution of (4) into conjugated boundary condition (2) yields the following expression

$$\sum_{k=1}^{n}\left\{2\,\mathrm{Re}(a_k H_k(\zeta))+\frac{d\zeta}{\overline{d\zeta}}\left(a_k\overline{\zeta}H_k'(\zeta)+b_k H_k(\zeta)\right)\right\}=N(\zeta)-iT(\zeta),\quad \zeta\in\Gamma \tag{5}$$

Unknown coefficients a_k and b_k can be determined from a system of linear algebraic equations obtained from (5) by diverse methods, for instance by collocations, which is accepted in the Trefftz approach.

2.2 Modification for cracks

For the case when contour Γ has sharp corners direct application of (5) is limited. This is explained by the presence of weak singularities in functions $H_k(z)$, which may lead to strong singularities in $H'_k(z)$. In fact, in the case of cracks some of the functions $H_k(z)$ (or all of them) have square root singularities at the crack tips and the derivatives will have singularities of order –3/2. Therefore, a modification of (5) is necessary to avoid the presence of unphysical singularities, which can be done in the following way. Firstly, we introduce a special holomorphic function $\chi(z)$ that nullifies the difference $\overline{\zeta}-\chi(\zeta)$ at all crack tips and other sharp corners. Then the stress deviator (second formula in (1)) can be presented in the form

$$D(z,\overline{z})=(\overline{z}-\chi(z))\Phi'(z)+\omega(z),\quad \omega(z)=\chi(z)\Phi'(z)+\Psi(z) \tag{6}$$

where $\omega(z)$ is a new unknown holomorphic function in Ω. We further seek this function in the form

$$\omega(z)=\sum_{k=1}^{n} a_{k+n} H_k(z) \tag{7}$$

Thus, the function $\omega(z)$ (and consequently the stress deviator) has no strong singularities at the crack ends and sharp corners, and hence all stress components can only have singularities not stronger than square root.

The notation for unknown coefficients in (7) has been introduced for convenience and allows one to represent the modified expression (5) in the following form

$$\sum_{k=1}^{n}\left\{2\,\mathrm{Re}\left(a_k H_k(\zeta)\right)+a_k\bar{\zeta}\frac{d\zeta}{d\bar{\zeta}}H_k'(\zeta)\right\}+ \\ +\sum_{k=n+1}^{2n} a_k\frac{d\zeta}{d\bar{\zeta}}H_{k-n}(\zeta)=N(\zeta)-iT(\zeta),\quad \zeta\in\Gamma \tag{8}$$

By introducing real coefficients

$$a_k=C_k+iC_{2n+k},\quad k=1\ldots 2n \tag{9}$$

one can rewrite (8) in the form

$$\sum_{k=1}^{4n}F_k(\zeta)C_k=p(\zeta),\quad \zeta\in\Gamma \tag{10}$$

Here the following notation has been introduced

$$F_k(\zeta)=\begin{cases} H_k(\zeta)+\overline{H_k(\zeta)}+\bar{\zeta}e^{2i\theta(\zeta)}H_k'(\zeta) & 1\le k\le n \\ i\left(H_k(\zeta)-\overline{H_k(\zeta)}-\bar{\zeta}e^{2i\theta(\zeta)}H_k'(\zeta)\right) & n<k\le 2n \\ e^{2i\theta(\zeta)}H_{k-n}(\zeta) & 2n<k\le 3n \\ ie^{2i\theta(\zeta)}H_{k-n}(\zeta) & 3n<k\le 4n \end{cases} \tag{11}$$

is the set of known functions expressed via basis functions $H_k(z)$;

$$p(\zeta)=N(\zeta)-iT(\zeta) \tag{12}$$

is known stress vector on the contour;

$$e^{2i\theta(\zeta)}=\frac{d\zeta}{d\bar{\zeta}},\quad \zeta\in\Gamma \tag{13}$$

where θ is the angle between the tangent to the contour at point ζ and real axis of a Cartesian coordinate system.

2.3 On independence of coefficients C_k

Unknown coefficients in (3) may be dependent. This is evident from Savruk's representation [8] of complex potentials via the Cauchy integrals

$$\Phi(z)=\frac{1}{2\pi}\int_\Gamma\frac{Q(t)}{t-z}dt,\quad \Psi(z)=\frac{1}{2\pi}\int_\Gamma\left\{\frac{\overline{Q(t)}-2i\overline{q(t)}}{t-z}\overline{dt}-\frac{\bar{t}Q(t)}{(t-z)^2}dt\right\} \tag{14}$$

where $2q(t)$ is a jump of $N+iT$ across the contour. If this jump vanishes on Γ then both complex potentials are expressed via the density of the crack opening displacements $Q(t)=2G(1+\kappa)^{-1}(u'^{+}_1-u'^{-}_1+i(u'^{+}_2-u'^{-}_2))$ alone. This function can be sought as a finite series, which after integration results in the following expressions

$$Q(t)=\sum_{k=1}^{n} a_k q_k(t),\quad \Phi(z)=\sum_{k=1}^{n} a_k H_k(z),\quad \Psi(z)=\sum_{k=1}^{n} \bar{a}_k G_k(z)-a_k T_k(z)$$
$$H_k(z)=\frac{1}{2\pi}\int_{\Gamma}\frac{q_k(t)}{t-z}dt,\quad G_k(z)=\frac{1}{2\pi}\int_{\Gamma}\frac{\overline{q_k(t)}}{t-z}d\bar{t},\quad T_k(z)=\frac{1}{2\pi}\int_{\Gamma}\frac{\bar{t}q_k(t)}{(t-z)^2}dt \tag{15}$$

Here the coefficients b_k are absent but different sets of basis functions is used for the representations of the second complex potential. The functions G_k(z) and T_k(z) can be expressed through H_k(z) if n tends to infinity, which will indicate dependence of the coefficients. If n is finite then, in general, the coefficients can be considered as independent, at least with certain accuracy. However, the coefficients b_k can be expressed through the coefficients a_k in some cases of simple contours. This becomes evident if one considers a straight crack in the plane, in this case the solution can be derived via the potential $\Phi(z)$ only [1], which indicates that all coefficients b_k in (3) depend on a_k. In complex cases of geometry this does not take place, however one cannot expect that all coefficients C_k will be independent of each other for particular problems. Complete investigation of independence of representations (3) and/or (7) presents essential difficulties and it is out of scope of this paper. It is further accepted that in numerical calculations any dependence of the coefficients will manifests itself in ill-condition matrix of linear algebraic system. Therefore, it necessitates application of a regularisation technique regardless of the method used for solution of (10).

2.4 Regularised approximate solution of the problem

Method of collocation is further used to solve (10). For this purpose, N nodes are selected on the boundary of the domain $\zeta_j \in \Gamma$, which results in the following linear algebraic system of equations

$$\sum_{k=1}^{4n} F_{kj}C_k = p_j,\quad j=1\ldots N;\quad F_{kj}=F_k(\zeta_j),\quad p=p(\zeta_j), \tag{16}$$

The coefficients of this system and the right hand side are both complex, which is not suitable for the SVD method used to solve (16). Therefore, the following system has been formed

$$\mathbf{MC}=\mathbf{B} \tag{17}$$

where **C** is $4n$-vector of unknowns with the components C_k, **M** is $2N$x$4n$ matrix of the system, M_{km}=Re(F_{kj}), m=1…N, M_{km}=Re(F_{kj}); m=N+1…2N., k=1…4n; **B** is $2N$-vector of applied load with the components B_m=Re(p_m), m=1…N; B_m=Im(p_m), m=N+1…2N.

The number of collocations is chosen to make system (17) overdetermined. Then the inverse matrix $\mathbf{M}^{-1}$ in (17) is found by the singular value decomposition, SVD (see e.g. Golub and van Loan [9]). This method allows to control the condition number of the matrix *Cond*(**M**) and if necessary to perform regularisation if *Cond*(**M**) is greater than a chosen threshold, C^*. The solution of (17) is presented as follows

$$\mathbf{C} = \mathbf{M}_{reg}^{-1}\mathbf{B}, \quad \mathbf{M}_{reg}^{-1} = \mathbf{V}^T\mathbf{D'}\mathbf{U} \tag{18}$$

Here **U** ($4n$x$2N$) and **V** ($4n$x$4n$) are orthogonal matrices in the SVD of the matrix **M**, $\mathbf{M}=\mathbf{U}\mathbf{D}\mathbf{V}^T$, **D** is ($4nx4n$) diagonal matrix formed from the singular values, d_j, of the matrix **M** placed in descending order, $d_1 \geq d_2 \geq \ldots \geq d_n$, **D′** is the diagonal matrix of the rank k as follows $\mathbf{D'}=diag\{d_1^{-1}, d_2^{-1}, \ldots d_k^{-1}, 0\ldots 0\}$. It should be noted that $\mathbf{D'}=\mathbf{D}^{-1}$ if no regularisation has been made.

3 Numerical examples

3.1 Crack in a plane

Let us consider an elastic isotropic plane with a crack lying on the interval ($-L,L$) of the real axis. Let the surfaces of the crack be subjected to normal and shear load $p=N+iT$ (Fig 1a). An analytical solution for this configuration can be found, in particular, stress intensity factors, SIFs, at the right crack tip are found in the following form [8]

$$K_I^{ideal} - iK_{II}^{ideal} = \frac{-1}{\sqrt{\pi L}} \int_{-L}^{L} \sqrt{\frac{L+t}{L-t}}\, p(t)dt \tag{19}$$

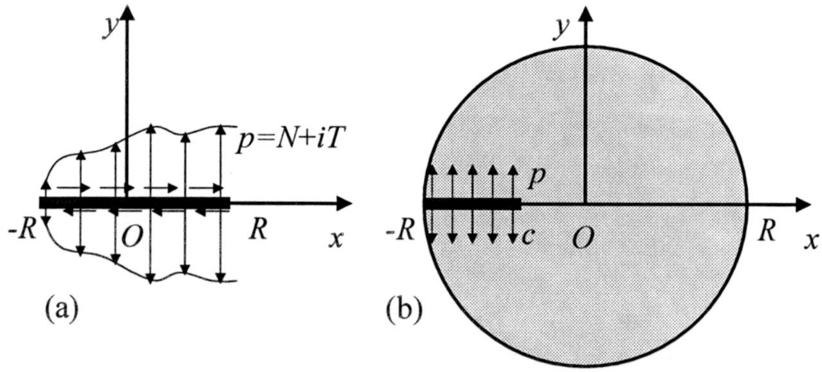

Figure 1: Crack in the plane (a), edge crack in the circle (b).

The following basis functions are used

$$H_k(z) = k\left(\sqrt{z^2 - L^2} - z\right)^{k-1}\left(\frac{1}{\sqrt{z^2 - L^2}} - 1\right) \tag{20}$$

These provide correct behaviour of sought solutions at infinity and at the crack tips, which is important because improper choice of basic functions may lead to incorrect solutions.

Calculated stress intensity factors are found in the form

$$\frac{K_I - iK_{II}}{p_0\sqrt{\pi L}} = \sum_{k=1}^{n}(-1)^{k-1}k\left(2\,\mathrm{Re}(a_k)+a_{n+k}\right)L^{k-1} \tag{21}$$

where p_0 is a parameter of loads.

Calculations have been performed for different polynomial loads, $p(t)=\alpha_m(x/L)^m$ with different n and N. The condition number has been kept below 10^7. The results for SIFs are accurate if $m<n$, which is explained by proper choice of basis functions (in this case $p_0=\alpha_m$). Tests have been also made for other load types and they show good agreement with analytical solutions. Thus solutions for step-like load on the crack have been checked. Fig 2 shown one example in which n=36, N=576, C^*=212 and calculates SIF has an error of less than 0.8%, $K_I^{\text{ideal}}=1/3\ (pL)^{1/2}$.

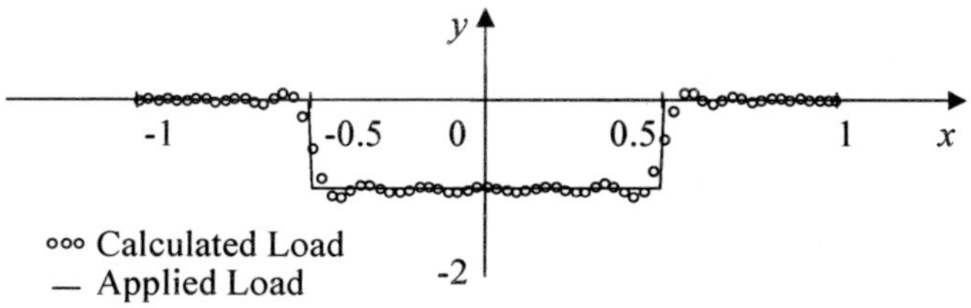

Figure 2: Boundary stresses for the case for step-like load applied to the crack in plane.

3.2 Edge crack in a circle

Let us consider a crack (of the length L) in an elastic circle of the radius R; the crack is located on the interval ($-R$,$-R+L$) of the real axis, Fig 2b. The surfaces of the crack are subjected to normal pressure p. An analytical solution for this configuration can be found, in particular, mode I stress intensity factor is found in the following form [8]

$$K_I^{ideal} = 3.17214(2-L)^{-\frac{3}{2}}\,p\sqrt{\pi L} \tag{22}$$

The following basis functions are used in modelling

$$H_k(z) = (z-c)^{\frac{k}{2}-1} \tag{23}$$

This selection provides correct behaviour of the solution at the crack tip and at the crack mouth.

Calculated mode I SIF is found as follows

$$K_I = \left(\frac{a_1}{2} + \overline{a}_1 + a_{n+1}\right)\sqrt{2\pi p} \tag{24}$$

Calculations have been performed for different crack lengths varying from $0.2R$ to $1.8R$. Different combination of parameters n and N have been checked and in all tests the threshold for the condition number has been set $C^* = 10^8$. The results for K_I are satisfactory for $0.2<L/R<1.4$ and the errors can be made within 1% by the proper selection of parameters n and N. For long cracks the numbers n and N should increase, which makes the condition number of the matrix to be quite big and therefore the SVD regularisation cannot produce accurate results. An example of boundary stress calculations is presented in Fig 3 for the case $L=1.2R$, where for $n=36$, $N=468$ (180 nodes have been selected on the circumference and $4n$ nodes on the upper and lower boundaries of the crack, i.e. 288 nodes in total the contour modelling the crack). The condition number of the matrix **M** is greater than 10^{15}, therefore the regularisation has been performed, $Cond(\mathrm{M}_{reg}^{-1})=6.85\cdot 10^7$. Despite reduced accuracy caused by regularisation the result for $K_I=4.437(\pi L)^{1/2}p$ is within 0.1% error.

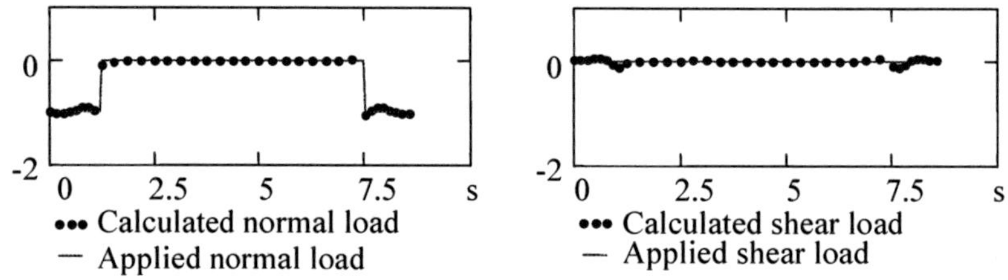

Figure 3: Normal (lefts) and shear (right) stresses on the contour in the case of the edge crack in the circle, $L=1.2R$.

4 Conclusions

It is shown that the Trefftz approach in which complex potentials are approximated by linear combinations of holomorphic functions can be used for solving fracture mechanics problems. It is important to ensure correct behaviour of the sought solutions at the crack tips and at the sharp corners. This has to be achieved by the proper selection of the holomorphic functions used for representation of the complex potentials. The method leads to an overspecified system of linear algebraic equations, which requires regularisation. The latter can be made on the basis of the SVD methods, which also ensures independence of the coefficients in truncated series for complex potentials. Two considered examples indicate that the proposed approach is capable to solve the problems with straight cracks in infinite and finite domains. The analysis of more complex geometries involving curvilinear and branching cracks is required in order to conclude applicability of the suggested approach for the general case.

References

[1] Muskhelishvili, N.I. Some basic problems of the mathematical theory of elasticity, P. Noordhoff, Groningen, the Netherlands, 1963.

[2] Zielinski, A.P., 1988. Treffz method: elastic and elastoplastic problems. Computer Methods In Applied Mechanics And Engineering, **69**, 185–204.

[3] Kita, E. and Kamiya, N., 1995. Trefftz method: an overview. Advances in Engineering Software, **24**, 3–12.

[4] Domingues, J.S., Portela, A. and. Castro, P.M.S.T., 1999. Trefftz boundary element method applied to fracture mechanics. Engineering Fracture Mechanics, **64**, 67–86.

[5] Li, Z-C., Lu, T-T. and Hu, H-Y., 2004. The collocation Trefftz method for biharmonic equations with crack singularities. Engineering Analysis with Boundary Elements, **28**,79–96.

[6] Galybin, A.N., 2005. Approximation of complex potentials as a uniform approach for solving classical and non-classical boundary value problems of plane elasticity. WIT Transactions on Modelling and Simulation, **39**, WIT Press, Southampton, UK, 255–264

[7] Galybin, A.N. and Mukhamediev, Sh.A. 2004. Determination of elastic stresses from discrete data on stress orientations. Int. Journal of Solids and Structures. **41** (18–19), 5125–5142.

[8] Savruk, M.P. Two-dimensional elastic problems for bodies with cracks. Naukova Dumka, Kiev, 1981.

[9] Golub, G.H. and van Loan C.F. Matrix computations, The Johns Hopkins Baltimore / London: University Press, 1989.

Integral transform approach for dynamic fracture calculation and crack growth modelling

P. Brož
Professional Association for Science, Research and Consultancy, Czech Republic

Abstract

The analytical mould of the solution in the radial direction permits precision stress intensity factors to be derived in the scaled boundary finite element method, namely, directly from the definition, and consequently no particular crack-tip interpretation such as refining the crack-tip mesh or employing singular elements, is indispensable. Besides, anisotropic material characteristics can be treated effortlessly. In this analysis, the Frobenius approach in the frequency domain to the solution of the governing differential equations of the SBFEM is used to simulate variable dynamic fracture problems. The complex frequency-response functions are calculated. Thereafter, the dynamic stress intensity factors are squarely taken from the response functions out. That is ensued by a fast Fourier transform of the transient load and a later inverse transform to derive the time history of DSIFs.

A mixed-mode crack growth simulation was developed. At first, a domain is divided into some subdomains. Since the dimensions and shapes of subdomains can be flexibly changed and only the domain boundaries or common edges between subdomains are discretized in the SBFEM, a remeshing routine such as a straightforward one as in BEMs was set up with minimum mesh variances while the universality and flexibleness of the FEM is preserved.

Keywords: crack trajectory, frequency-domain approach, response function, internal damping coefficient, load-displacement relation, mesh density, remeshing routine, subdomain, transient fracture problem.

WIT Transactions on Modelling and Simulation, Vol 47, © 2008 WIT Press
www.witpress.com, ISSN 1743-355X (on-line)
doi:10.2495/BE080151

1 Introduction

For analyzing dynamic stress intensity factors, several numerical methods were evolved and used. As regards how the simulated domain is discretized in space, these techniques may be in general split up into four types: the finite difference method (FDM), the finite element method (FEM), the boundary element method (BEM); the meshless or meshfree method are more recently and extensively used in modelling mixed-mode crack growth. As regards how the time processes of DSIFs are calculated, the methods given can be arranged into time-domain ways based on the direct time integration, and integral transform methods inclusive of the Laplace transform and the Fourier one (that means, the frequency-domain method in virtue of frequency analysis, and the discrete or fast Fourier transform (FFT). Superb literature surveys of these methods were presented by Ariza and Dominguez [1], and Song [2].

A frequency-domain interpretation approach has certain benefits over the time-domain, such as no demand for a mass matrix, and so coarser meshes may be applied, and as soon as a complex frequency-response function is obtained, it can be employed in combination with FFT and inverse FFT (IFFT) to compute transient responses for different forms of dynamic loading.

The scaled boundary finite-element method (SBFEM), developed lately in [3,4], is a semi-analytical technique joining the advantages of FEM and BEM, i.e., it discretizes boundaries merely so that the simulated spatial dimensions are made smaller by one like the BEM and meantime it does not require fundamental solutions as the FEM.

In consequence, the considerable usability of the FEM and the easiness in remeshing of the BEM can be potentially kept.

The modelling of mixed/mode crack propagation is accomplished by combining the remeshing routine with a growth criterion. The developed simulation is competent of predicting crack trajectories and load-displacements dependences precisely and effectively.

Model problems with isotropic and anisotropic material characteristics are simulated by applying the frequency-domain procedure. The influences of the internal damping coefficient, the mesh density, the frequency interval and the maximum frequency are debated as well. The DSIFs K_I and K_{II} are extracted from the stress solution for a wide range of frequencies, resulting in complex frequency-DSIFs response functions. These functions are later applied to generate time histories of the DSIFs. This procedure is accurate and efficient with a small number of degrees of freedom.

In the paper, a frequency-domain technique is evolved for transient dynamic fracture analysis applying the SBFEM according to [5]. The Frobenius solution [6] developed in a different way in the frequency domain, is used to determine complex frequency-DSIFs functions that are afterwards employed with fast Fourier transform and inverse FFT to calculate the time process of DSIFs.

The static conception within the bounds of the SBFEM is presented, including instructions how to extract SIFs from the solution, in compliance with [6]. Considering the advantages of the SBFEM, a very straightforward remeshing routine is evolved, in company with a concise discussion of mixed-mode crack growth criterion and explanation of the scaling routine.

2 Computation of dynamic stress intensity factors

A cracked region simulated by the SBFEM is indicated in fig. 1. All the time, the scaling center is located at the crack tip. The square-root singularity existing at this tip in a homogenous plate are the circumstances very considerably studied. In this instance the mode-I and mode-II SIFs are defined being

$$\begin{Bmatrix} K_I \\ K_{II} \end{Bmatrix} = \lim_{r\to 0} \begin{Bmatrix} \sqrt{2\pi r}\,\sigma_{yy}\big|_{\theta=0} \\ \sqrt{2\pi r}\,\sigma_{xy}\big|_{\theta=0} \end{Bmatrix} \tag{1}$$

where r and θ are the polar coordinates with the origin at the crack tip (see fig.1).

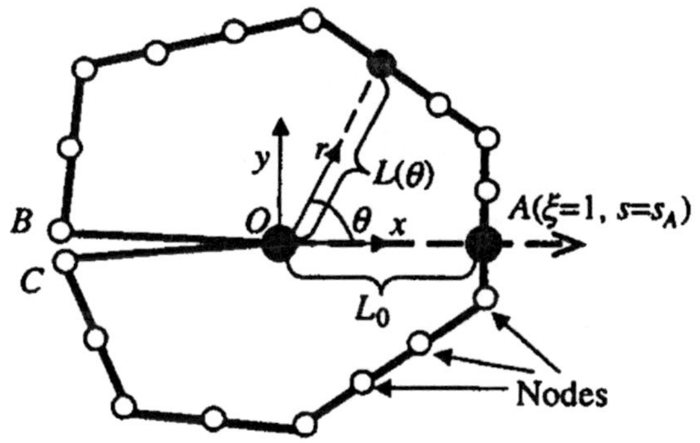

Figure 1: A cracked domain simulated by the scaled boundary finite-element method (taken from [5]).

Notice the relation between r and ξ (ξ =0 at the crack tip and 1 at the boundary, according to fig. 1) has the form

$$r = \xi L(\theta) \tag{2}$$

where $L(\theta)$ is the distance between the crack tip and the intersection point of the polar line r and the region boundary. In compliance with the Frobenius solution procedure outline and considering Eq. (1) we get

$$\begin{Bmatrix} K_I \\ K_{II} \end{Bmatrix} = \lim_{r\to 0} \begin{Bmatrix} \sqrt{2\pi L(\theta)}\sum_{i=1}^{n} c_i \xi^{(\lambda_i - 0.5)} \psi_{yy}(s)_i\big|_{\theta=0} \\ \sqrt{2\pi L(\theta)}\sum_{i=1}^{n} c_i \xi^{(\lambda_i - 0.5)} \psi_{xy}(s)_i\big|_{\theta=0} \end{Bmatrix} \tag{3}$$

The stress field expression is a sign of all the stress modes with $\lambda_i \geq 1$, disappears for $\xi \to \theta$ or $r \to 0$. Just two modes with $\lambda_i = 0.5$ results in singular stresses for $\xi \to \theta$ or $r \to 0$. In view of both dynamic stress solution and its all added summation terms possessing ${}^{m}\lambda_i \geq 2$ $(2 \leq m \leq k+1)$, e.g. is universal for static and as dynamic problems. After indicating these two stress modes as mode

I and mode II for λ_i = 0.5, and determining the limit in Eq. (3), it ends in the following SIFs:

$$\begin{Bmatrix} K_I \\ K_{II} \end{Bmatrix} = \begin{Bmatrix} \sum_{i=\mathrm{I,II}} \sqrt{2\pi L(\theta)} c_i \psi_{yy}(s)_i \big|_{\theta=0} \\ \sum_{i=\mathrm{I,II}} \sqrt{2\pi L(\theta)} c_i \psi_{xy}(s)_i \big|_{\theta=0} \end{Bmatrix} = \sqrt{2\pi L_0} \sum_{i=\mathrm{I,II}} \left(c_i \begin{Bmatrix} \psi_{yy}(s = s_A) \\ \psi_{xy}(s = s_A) \end{Bmatrix}_i \right) \tag{4}$$

Where $L_0 = L(\theta = 0)$, what means the distance between the crack tip and the point A($\xi = 1$, $s = s_A$) in the direction of crack surface on the boundary (fig.1) The above definition is based on the local coordinate system where the cracking direction is supposed to coincide with the global x-axis.

For a domain involving a crack surface at an arbitrary way, the stresses should first be changed by the standard method to normal (mode I) stress σ_n and shear (mode II) stress τ_n on the cracking surface plane at the point A so that Eq. (4) may be considered:

$$\begin{Bmatrix} K_I \\ K_{II} \end{Bmatrix} = \sqrt{2\pi L_0} \sum_{i=\mathrm{I,II}} \left(c_i \begin{Bmatrix} \sigma_n(s = s_A) \\ \tau_n(s = s_A) \end{Bmatrix}_i \right) \tag{5}$$

It should be remarked that the point A does not need to be a prevailing node.

3 Fast Fourier transform to quantify transient response

Regard a periodic excitation force $p(t)$ by N number of discrete values explained $p_n = p(t_n) = p(n\Delta t)$ where $\Delta t = T_0 / N$ is the sampling interval, T_0 the period and n extends from 0 to N-1. The array p_n characterizing the discretized force function may be rendered being a superposition of N harmonic functions

$$p_n = \sum_{j=0}^{N-1} P_j e^{i(j2\pi f_0 t_n)} = \sum_{j=0}^{N-1} P_j e^{i(2\pi nj/N)}, \quad n = 0,\dots, N-1 \tag{6}$$

where $f_0 = 1/T_0$ stands for frequency (Hz) of first harmonic in the periodic extension of $p(t)$, $f_j = j.f_0$ is the frequency of the jth harmonic, and P_j is a complex-valued coefficient which determines the amplitude and period of jth harmonic. P_j can be written as

$$P_j = \frac{1}{N} \sum_{j=0}^{N-1} P_j e^{-i(j2\pi f_0 t_n)} = \frac{1}{N} \sum_{j=0}^{N-1} P_j e^{-i(2\pi nj/N)}, \quad j = 0,\dots, N-1 \tag{7}$$

Eqs (6) and (7) circumscribe a discrete Fourier transform two: the array P_j is the DFT of the excitation sequence p_n, and the array p_n means the inverse of DFT (IDFT) of the sequence P_j. The frequency of the highest harmonic comprised in Eqs (6) and (7) is known being Nyquist frequency, introduced by $N.f_0/2 = 1/(2\Delta t)$.

A complex frequency-response function H_j is calculated for each $f = f_j$ ($0 \le j \le N/2$). Taking into account that Eq. (6) is a one-sided Fourier expansion, the values on both sides of $j = N/2$ have to be complex conjugates of each other, that means

$$H_j = H^*_{N-j}, \quad j = \frac{N}{2} + 1,\dots, N-1 \tag{8}$$

where H^* indicates the complex conjugate of H. The response to each harmonic constituent of the excitation can be evaluated in the form

$$U_j = H_j P_j, \quad j = 0,...,N-1 \tag{9}$$

Lastly, the response $u_n = u(t_n)$ at discrete time moments $t_n = n\Delta t$ is calculated by the IDFT of U_j

$$u_n = \sum_{j=0}^{N-1} U_j e^{i(j2\pi f_0 t_n)} = \sum_{j=0}^{N-1} U_j e^{i(2\pi nj/N)}, \quad n = 0,...,N-1 \tag{10}$$

The aforesaid passage deals with the classical DFT solution to the transient elastodynamics, namely in the frequency domain. In so doing, the fast Fourier transform is a very effective and precision way of the DFT - IDFT two determination.

4 Questions under discussion

To prove the evolved frequency-domain technique based on the SBFEM, one standard dynamic fracture problem was simulated. Concurrently, complex frequency response functions for frequencies f_j(j = 1,…, M) are reckoned using the Frobenins solution procedure. The minimum frequency is f_1 = 0 Hz (static case) and the maximum frequency is represented by the quantity $f_M = f_{max}$. At the same time, the damping effect is taken into consideration by adapting the elastic moduli to include an internal damping coefficient β, charactering the complex Young's modulus $E_c = E(1 + i2\beta)$ and the complex shear modulus $G_c = G(1 + i2\beta)$ where E and G are the elastic Young's modulus and the elastic shear modulus, respectively. The dynamic stress intensity factors K_{Ij} and K_{IIj} (j = 1,…, M) are extracted next directly from the stress responses. The foregoing operation is succeeded by a FFT of the transient load Eq. 7 to get the coefficients P_j (j = 0,..., N-1). The frequency responses H_j (j = 0,…, N/2) are then interpolated from the M number of K_{Ij} and K_{IIj}. As a rule, the interpolation is indispensable for a much greater N (an integer power of 2) can be desirable to precisely meant the transient load while the complex response functions may be quite example of a smaller M number of frequencies. This may put by extensively computing cost. The proper frequency-response functions are made whole by Eq. (8). Lastly, an inverse FFT Eq. (10) of U_j (j=0, …., N) calculated by Eq. (9) is performed to obtain the response u_n (n = 0, …, N), that is, the time process of dynamic stress intensity factors. The functions *fft*( ) and *ifft*( ) are employed according to MATLAB to conduct FFT and IFFT, respectively. In so doing, damping coefficients of β = 0.0 , 0.001, 0.01, 0.025 and 0.05 are simulated. For each β, the excitation frequencies f_1 = 0 to f_{max} = 500 000 Hz with an interval Δf = 1000 Hz are modelled, ending in M = 501 points on a complex frequency response curve. A convergence tolerance of $\alpha = 1e-3$ is applied for all the analyses. The influences of the mesh density, the damping coefficient β, the maximum frequency f_{max} and the frequency interval $\Delta f = f_{max}/(M\text{-}1)$ on the final exactitude and the computational cost are treated. For all the calculations, two-node linear line elements are applied.

5 Rectangular plate weakened by a slanting edge crack

The said isotropic plate is subject to uniform traction on the upper surface. The geometry of the plate and boundary condition are demonstrated in fig. 2a.

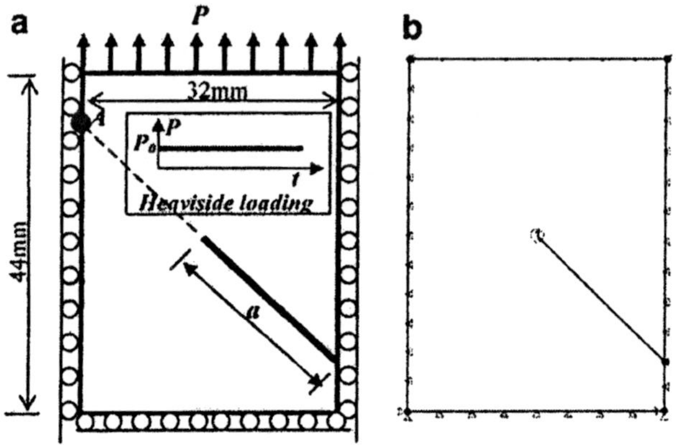

Figure 2: A rectangular plate with a slanting edge crack. (a) Dimensions, boundary and loading conditions, (b) coarse mesh including.

The material properties mean the shear modulus G = 29 GPa, Poisson's ratio υ = 0.286 and density ρ = 2450 kg/m^3. A mixed-mode fracture problem is the matter for which the Heaviside step loading is simulated (see fig. 2a). At the same time, all SIFs are normalized by $P_0(\pi a)^{1/2}$ where P_0 stands for the quantity of the transient loading and a = 23 mm is the crack length, when supposing a plane strain condition. One subdomain is introduced possessing the scaling centre at the crack tip. One mesh with 41 nodes is indicated in fig. 2b. The stresses at the point A in fig (2.a) are used to extract SIFs in compliance with Eq. (5).

Examples of the calculated DSIFs are demonstrated in figs. 3 and 4 where the time histories of K_{I} and K_{II} are shown, computed employing the course mesh with varied material damping coefficients β compared to the issues obtained in by Fedelinski et al. using the BEM.

There is very fair material harmony applying β = 0.01 for both K_{I} and K_{II}. The coarse mesh and the fine mesh results in almost identical outcomes demonstrating that the course mesh with only 41 nodes is fine plenty (see fig. 4). The same isotropic problem with a little various material properties was simulated by Song [2] applying the time-domain SBFEM and a finer mesh with 23 super-elements, and over 200 nodes. This expresses one of the disadvantages of time-domain methods, that means a mass matrix has to be applied in time-integration. Since the mass matrix used in [2] is the low-frequency expansion of the dynamic stiffness matrix, it may just symbolize the inertial results at low frequencies. The higher frequency constituents have to be an example of fine meshes with small dimensions. The frequency-domain method, on the other

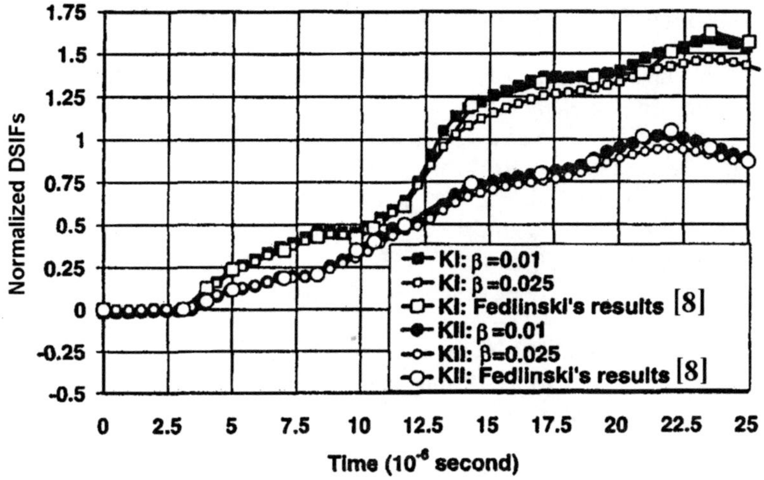

Figure 3: Normalized DSIFs applying the coarse mesh: influence of material damping coefficient β (taken from [5]).

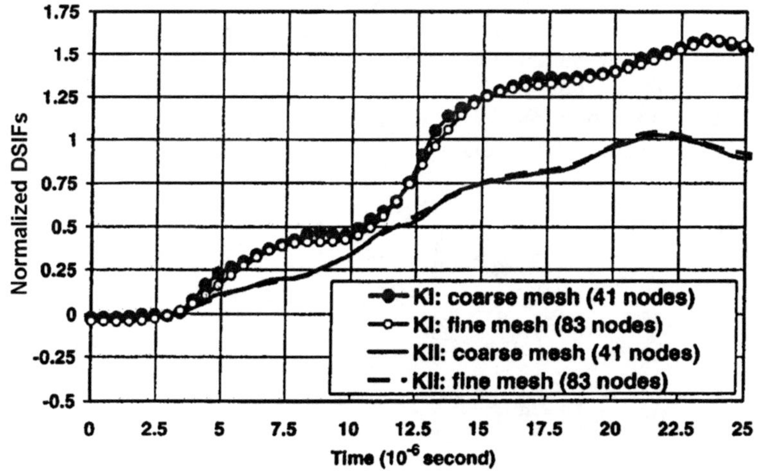

Figure 4: Normalized DSIFs using $\beta = 0.01$: influence of mesh density.

hand, does not require mass matrices. This permits precise computation of time responses applying coarse meshes like this as fig. 2(b). The orthotropic instance of the same plate (see fig. 2.(a)) in plane stress is modelled applying the identical mesh (fig. 2(b)). The properties in the principal material axes stand for: Young's modulus E_1 = 82 GPa, $E_2 = \mathrm{R} \cdot \mathrm{E}_1$, shear modulus G_{12} = 29 GPa, Poisson's ratio $\upsilon_{12} = 0.4006$ and the mass density $\rho = 2450$ kg/m^3. The same problem was solved by Song applying the time-domain SBFEM [2] and Albuquerque et al. using the BEM [7]. Two events of $R = 0.5$ and 2.0 are studied.

6 Methodology of simulations and remeshing routine

A domain can be advantageously divided into subdomains in the SBFEM just as it can be stood for by elements in the FEM. On account of the semi-analytical essence of the SBFEM, the dimension and shape of subdomains in a region can be altered to a great degree if desired without relinquishment of the resulting exactitude. As well, the scaling center can be put at the required place within the subdomain (if the domain boundary is observable from the scaling center). The nodal density in various subdomains can be very altered, too. A numerical model of LEFM – based mixed-mode crack growth applying the SBFEM presupposes the phases:

1. Subdivide the initial domain into some subdomains by straight lines or edges (see fig. 5) in view of the displacement constraints and loading conditions. The topological information of the domain is fully represented by the edges and their intersections, i.e., vertices.

Each edge is related with two vertices. Different seeds can be specified to different edges or subdomains so that the nodal density from edge to edge or from subdomain to subdomain can be flexibly varied as required. The loading conditions and boundary constrains are also related to edges: Each subdomain is fully represented by the connected vertices placed in a line counterclockwise round the scaling centre. Fig. 5 shows that a domain is composed of four subdomains with the subdomain 4 (S4) being the crack subdomain. Placing scaling centers for the normal subdomains can be optimized by taking advantage of non-discretization of unrestricted side-front surfaces.

Namely, we get the crack growth direction θ_0 from the conditions

$$\left.\begin{aligned} G(\theta) &= \frac{4}{E}\left(\frac{1}{3+\cos^2\theta}\right)^2\left(\frac{1+\frac{\theta}{\pi}}{1-\frac{\theta}{\pi}}\right)^{-\frac{\theta}{\pi}}\left[\left(1+3\cos^2\theta\right)K_I^2 - 8\sin\theta\cos\theta K_I K_{II}\right. + \\ &\qquad \left. + \left(9-5\cos^2\theta\right)K_{II}^2\right] \\ &\frac{\partial G(\theta)}{\partial\theta} = 0, \quad \text{and} \quad \frac{\partial^2 G(\theta)}{\partial\theta^2} < 0 \end{aligned}\right\} \tag{11}$$

For the maximum stress energy release rate it holds $G_{\max} = G(\theta)$. The scaling factor is equal to

$$\alpha = \sqrt{\frac{G_{IC}}{G_{\max}}} \tag{12}$$

where G_{IC} is the critical SERR or else fracture energy being interrelated to the material fracture toughness K_{IC} in the form

$$G_{IC} = \frac{K_{IC}^2}{E} \text{ for plane stress and } G_{IC} = \frac{\left(1-\nu^2\right)K_{IC}^2}{E} \text{ for plane strain} \tag{13}$$

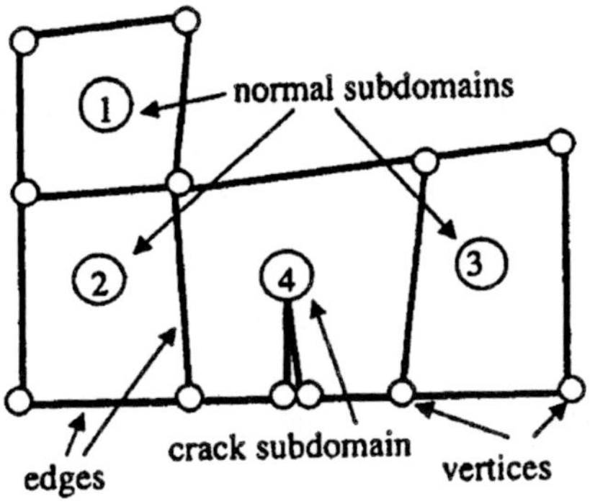

Figure 5: Subdomains.

Scaling the reference external loading vector and displacement reactions on crack growth is performed; it follows that

$$p_s = \alpha p_0$$
$$u_s = \alpha u_0 \tag{14}$$

where p_s is the scaled loading vector minimal for the crack to growth and u_s means the displacement field of the domain subject to p_s, with the stress field recovered in the region.

The easy remeshing routine permits fully automatic crack growth modelling to be attained without user interferences. Improvements may lead to more sophisticated remeshing procedures but the current one seems sufficient for numerical examples. The smoothness of modelled crack trajectories depends on the pre-determined crack addition length.

7 Conclusion

Investigations employing the developed frequency-domain way of tackling have been performed on two standard problems with both isotropic and orthotropic material characteristics. Numerical issues indicate that this approach is competent of calculating time processes of DSIFs properly and strikingly involving a small number of degrees of freedom. It is expected when combined with the simple remeshing in the procedure SBFEM, this frequency-domain way will provide a very competitive system for simulating dynamic crack growth tasks. As well, it is learned that calculating the complex frequency-response functions is considerably time-consuming than FFT and IFFT.

The maximum frequency and the frequency interval that are important factors for the complex frequency-response functions to be applied in IFFT, should be chosen suitably. Internal material damping is considered to influence numerical both outcomes and stability.

The newly-evolved SBFEM was used to simulate the LEFM-based mixed-mode crack growth in brittle and quasi-brittle materials.

Acknowledgements

The author gratefully acknowledges the financial support of the presented research by the Grant Agency of the Czech Republic (project No. 103/06/1382) and Grant Agency of the Academy of Sciences of the Czech Republic (project IAA 200710604).

References

[1] Ariza, M.P., Dominguez, J., General BE approach for three-dimensional dynamic fracture analysis. *Engng Anal Bound Elem*, **26**, pp. 639–651, 2002.
[2] Song, C.M., A super-element for crack analysis in the time domain. *Int J. Numer Meth Engng*, **61**, pp. 1332–1357, 2004.
[3] Wolf, J.P., Song C.M., *Finite-element modelling of unbounded media.* Chichester: John Wiley and Sons, 1996.
[4] Wolf J.P., *The scaled boundary finite element method.* Chichester: John Wiley and Sons, 2003.
[5] Yang, Z.J., Deeks, A.J., HaO, H.A., Transient dynamic fracture analysis using scaled boundary finite element method: a frequency-domain approach. *Engineering Fracture Mechanics*, **74**, pp. 669–687, 2007.
[6] Yang, Z.J., Fully automatic modelling of mixed-mode crack propagation using scaled boundary finite element method. *Engineering Fracture Mechanics,* **73**, pp. 1711–1731, 2006.
[7] Albuquerque, E.L., Sollero, P., Aliabadi, M.H., The boundary element method applied to time dependent problems in anisotropic materials. *International Journal of Solids and Structures*, **39**, pp. 1405–1422, 2002.

Section 5
Mesh reduction techniques

Desingularised method of double layer fundamental solutions for potential flow problems

B. Šarler
Laboratory for Multiphase Processes,
University of Nova Gorica, Slovenia

Abstract

This paper describes an application of the recently proposed Modified Method of Fundamental Solutions (MMFS) to potential flow problems. The solution in two dimensional Cartesian coordinates is represented in terms of the double layer fundamental solution of the Laplace equation. The collocation is used for determination of the expansion coefficients. This novel method does not require a fictitious boundary unlike the conventional Method of Fundamental Solutions (MFS). The source and collocation points thus coincide on the physical boundary of the system. The desingularised value of the fundamental solution is in case of the coincidence of the collocation and source points determined from the respective integral equations logic. The values of the derivatives of the double layer fundamental solution in the coordinate directions, as required in potential flow calculations, are calculated indirectly from the considerations of the constant potential field. A numerical example of potential flow around the two dimensional circular region is shown. The results with the MMFS are compared with the results of the classical single layer MFS with an artificial boundary, and the analytical solution. It is shown that the MMFS gives better accuracy of the velocity components as compared with the classical MFS.

Keywords: potential flow, method of fundamental solutions, desingularisation, double layer potential.

1 Introduction

In recent years there has been strong development in mesh reduction methods in which the polygon-like meshes are reduced or avoided. The MFS is a

WIT Transactions on Modelling and Simulation, Vol 47, © 2008 WIT Press
www.witpress.com, ISSN 1743-355X (on-line)
doi:10.2495/BE080161

numerical technique that falls into the class of methods generally called boundary methods. The other well known representative of these methods is the Boundary Element Method (BEM) [1]. Both methods are best applicable in situations where a fundamental solution to the partial differential equation in question is known. In such cases, the dimensionality of the discretization is reduced. BEM for example requires polygonisation of the boundary surfaces in general 3D cases, and boundary curves in general 2D cases. This method requires solution of the complicated regular, weakly singular, strongly singular, and hypersingular integrals over boundary segments, which is usually a cumbersome and non-trivial task. The MFS has certain advantages over BEM, which are mostly visible in the fact that only pointisation of the boundary is needed, completely avoiding any integral evaluations, and making no principal difference in coding between the 2D and the 3D cases. A comprehensive survey of the MFS and related methods for elliptic boundary value problems can be found in [2–4]. In the present paper, the potential flow problem, previously solved by the least squares version of MFS [5] and collocation version of MFS [6] is solved by the MMFS. This novel method, which essentially represents a sort of blend between BEM and MFS, was originally developed by Young et al. [7] using collocation with the double layer Laplace equation fundamental solution. The method has been further extended to the single-layer Laplace equation fundamental solution in [8]. The main drawback of the MFS is representation of positions of the source points that need to be positioned outside the boundary. In cases where they are too close to the boundary, the solution is not accurate. In cases where they are too far away from the boundary, the discretisation matrix becomes ill conditioned. The novel MMFS overcomes this difficulty by allowing the source point positions to coincide with the collocation points on the physical boundary. A desingularisation technique thus has to be employed in order to be able to allow bounded values in the discretisation matrix. The desingularisation has been derived through the properties of the double layer potential in [7] and through the indirect BEM formulation in [8]. In the present paper, the desingularisation is extended to the calculation of the desingularised values through the direct BEM approach, as well as calculation of the desingularised value of the partial (not normal) derivatives on the boundary, which was not the case in the previous two cited MMFS pioneering papers by D. L. Young's group.

2 Governing equations

Consider a connected two-dimensional domain Ω with boundary Γ. The domain is filled by a fluid that undergoes potential flow. The boundary is divided into the part Γ^E, which represents the external boundaries of the system and into the part Γ^I, which represents the internal boundary of the system i.e. $\Gamma = \Gamma^E \cup \Gamma^I$. The potential Φ is governed by the following boundary value problem: Laplace equation

$$\nabla^2 \Phi = 0, \tag{1}$$

and boundary conditions of the Dirichlet and Neumann type, located at the Dirichlet Γ^D and Neumann Γ^N parts of the boundary Γ, i.e. $\Gamma = \Gamma^D \cup \Gamma^N$

$$\Phi(\mathbf{p}) = \bar{\Phi}^D(\mathbf{p}), \quad \mathbf{p} \in \Gamma^D, \quad \frac{\partial \Phi}{\partial \mathbf{n}_\Gamma}(\mathbf{p}) = \bar{\Phi}^N(\mathbf{p}), \quad \mathbf{p} \in \Gamma^N, \tag{2,3}$$

with $\mathbf{p}$ standing for the position vector and $\mathbf{n}_\Gamma$ for the outward normal on the boundary Γ. $\bar{\Phi}^D$ and $\bar{\Phi}^N$ representing the Dirichlet and Neumann boundary condition forcing functions. Let us introduce a two dimensional Cartesian coordinate system with ortho-normal base vectors $\mathbf{i}_x$ and $\mathbf{i}_y$ and coordinates p_x and p_y, i.e. $\mathbf{p} = p_x \mathbf{i}_x + p_y \mathbf{i}_y$. The potential field velocity components are calculated from the potential Φ as

$$v_\xi(\mathbf{p}) = \frac{\partial \Phi}{\partial p_\xi}(\mathbf{p}); \ \xi = x, y. \tag{4}$$

It is the purpose of this paper to determine the steady state potential flow components as a function of the posed geometry, governing equation and boundary conditions.

3 Solution procedure

3.1 Solution of the potential flow

The common points of the MFS and MMFS for solution of the potential flow field are elaborated first. The differences are elaborated afterwards. The solution of the potential Φ is represented by the N_Γ global approximation functions $\psi_n(\mathbf{p})$ and their coefficients α_n

$$\Phi(\mathbf{p}) \approx \sum_{n=1}^{N_\Gamma} \psi_n(\mathbf{p}) \alpha_n. \tag{5}$$

The global approximation functions have the property

$$\nabla^2 \psi_n(\mathbf{p}) = \begin{cases} 0; \mathbf{p} \neq \mathbf{p}_n \\ \delta(\mathbf{p}_n); \mathbf{p} = \mathbf{p}_n \end{cases}; n = 1, 2, ..., N_\Gamma, \tag{6}$$

i.e., they are fundamental solutions of the Laplace operator. δ denotes the Kronecker symbol. For two dimensional problems in Cartesian coordinates, the fundamental solution equals to

$$\psi_n^*(\mathbf{p}) = \frac{1}{2\pi} \log \frac{r^*}{r_n}; \ r_n^2 = \mathbf{r}_n \cdot \mathbf{r}_n, \ \mathbf{r}_n = \mathbf{p} - \mathbf{s}_n = (p_x - s_{nx})\mathbf{i}_x + (p_y - s_{ny})\mathbf{i}_y, \tag{7}$$

where r^* denotes the reference radius and p_{0x} and p_{0y} represent the mean coordinates of the $\Gamma \cup \Omega$. The double layer fundamental solution, defined by

$$\psi_n^\diamond(\mathbf{p}) = \frac{\partial \psi^*}{\partial \mathbf{n}_s}(\mathbf{p}) = \frac{1}{2\pi} \left[\frac{(p_x - s_{nx}) n_{sx}}{(p_x - s_{nx})^2 + (p_y - s_{ny})^2} + \frac{(p_y - s_{ny}) n_{sy}}{(p_x - s_{nx})^2 + (p_y - s_{ny})^2} \right] \tag{8}$$

is used in the present paper. Let us introduce the boundary condition indicators in order to be able to represent the boundary collocation equations in a compact form. The Dirichlet χ^D and Neumann χ^N type of boundary conditions indicators are

$$\chi^D(\mathbf{p}) = \begin{cases} 1; \mathbf{p} \in \Gamma^D \\ 0; \mathbf{p} \notin \Gamma^D \end{cases}, \quad \chi^N(\mathbf{p}) = \begin{cases} 1; \mathbf{p} \in \Gamma^N \\ 0; \mathbf{p} \notin \Gamma^N \end{cases}. \tag{9,10}$$

The coefficients are calculated from a system of N_Γ algebraic equations

$$\sum_{n=1}^{N_\Gamma} \Psi_{jn}\alpha_n = b_j; j = 1,2,...,N_\Gamma. \tag{11}$$

$$\Psi_{jn} = \chi^D(\mathbf{p}_j)\psi_n(\mathbf{p}_j) + \chi^N(\mathbf{p}_j)\frac{\partial \psi_n}{\partial \mathbf{n}_\Gamma}(\mathbf{p}_j), \tag{12}$$

$$b_j = \chi^D(\mathbf{p}_j)\bar{\Phi}^D(\mathbf{p}_j) + \chi^N(\mathbf{p}_j)\frac{\partial \bar{\Phi}^N}{\partial \mathbf{n}_\Gamma}(\mathbf{p}_j), \tag{13}$$

The coefficients α_n can be expressed through inversion of the system (10), which gives

$$\alpha_n = \sum_{j=1}^{N_\Gamma} \Psi_{nj}^{-1}\left[\chi^D(\mathbf{p}_j)\bar{\Phi}^D(\mathbf{p}_j) + \chi^N(\mathbf{p}_j)\bar{\Phi}^N(\mathbf{p}_j)\right]; \quad n = 1,2,...,N_\Gamma. \tag{14}$$

The velocity field components are calculated as

$$v_\xi(\mathbf{p}) = \sum_{j=1}^{N_\Gamma} \frac{\partial \psi_n}{\partial p_\xi}(\mathbf{p})\alpha_n; \ \xi = x, y. \tag{15}$$

3.2 Classical method of fundamental solutions

The fundamental solution source points are located outside physical boundary, i.e. $\mathbf{p}_j \neq \mathbf{s}_j$ and $\mathbf{s}_j \notin \Omega$ in the classical MFS. One can consider that they form an artificial boundary. The proper location of the source points is not a trivial task. It can be observed that the accuracy improves with the increasing distance from the physical boundary up to some extent. However, the collocation matrices become increasingly ill conditioned with increased distance from the boundary. The explicit form of the partial derivatives of the fundamental solution are

$$\frac{\partial \psi^\lozenge}{\partial p_x}(\mathbf{p}) = -\frac{1}{\pi}\frac{(p_x - s_{nx})(p_y - s_{ny})n_{sy}}{\left[(p_x - s_{nx})^2 + (p_y - s_{ny})^2\right]^2}$$

$$+\frac{1}{2\pi}\left[\frac{1}{(p_x - s_{nx})^2 + (p_y - s_{ny})^2} - \frac{2(p_x - s_{nx})(p_x - s_{nx})}{\left[(p_x - s_{nx})^2 + (p_y - s_{ny})^2\right]^2}\right]n_{sx} \tag{16}$$

$$\frac{\partial \psi^{\diamond}}{\partial p_y}(\mathbf{p}) = -\frac{1}{\pi}\frac{(p_y - s_{ny})(p_x - s_{nx})n_{sy}}{\left[(p_x - s_{nx})^2 + (p_y - s_{ny})^2\right]^2}$$

$$+\frac{1}{2\pi}\left[\frac{1}{(p_x - s_{nx})^2 + (p_y - s_{ny})^2} - \frac{2(p_y - s_{ny})(p_y - s_{ny})}{\left[(p_x - s_{nx})^2 + (p_y - s_{ny})^2\right]^2}\right]n_{sx} \quad (17)$$

3.3 Modified method of fundamental solutions

The key point of the modified method of fundamental solutions represents desingularisation of the value of the fundamental solution, because in this case the source and the collocation points coincide, i.e. $\mathbf{p}_j = \mathbf{s}_j$. The desingularisation value can be directly set from the

$$\int_{\Gamma} \Phi \nabla \psi^* \cdot d\mathbf{\Gamma} - \int_{\Gamma} \psi^* \nabla \Phi \cdot d\mathbf{\Gamma} + c(\mathbf{s})\Phi(\mathbf{s}) = 0 \,. \quad (18)$$

The derivation of the upper equation with respect to $\mathbf{n}_s$ gives

$$\int_{\Gamma} \Phi \nabla \psi^{\diamond} \cdot d\mathbf{\Gamma} - \int_{\Gamma} \psi^{\diamond} \nabla \Phi \cdot d\mathbf{\Gamma} + c(\mathbf{s})\frac{\partial \Phi}{\partial \mathbf{n}_s}(\mathbf{s}) = 0 \,. \quad (19)$$

In case of the constant potential Φ, the equation (18) reduces to

$$\int_{\Gamma} \Phi \nabla \psi^* \cdot d\mathbf{\Gamma} + c(\mathbf{s})\Phi(\mathbf{s}) = -\int_{\Gamma} \Phi \psi^{\diamond} d\Gamma + c(\mathbf{s})\Phi(\mathbf{s}) = 0 \,, \quad (20)$$

and the equation (19) into

$$\int_{\Gamma} \nabla \psi^{\diamond} \cdot d\mathbf{\Gamma} = 0 \,. \quad (21)$$

The calculation of the desingularised value of the double layer fundamental solution can be performed through the discretisation of the Equation (20) in the following simple way

$$-\sum_{n=1}^{N_\Gamma} \frac{\ell_{n-} + \ell_{n+}}{2} \psi_j^{\diamond}(\mathbf{p}_n) + c_s(\mathbf{s}_j) = 0 \,, \quad (22)$$

$$\psi_j^{\diamond}(\mathbf{p}_j) = \frac{1}{\ell_{j-} + \ell_{j+}}\left[1 - \sum_{\substack{n=1 \\ n \neq j}}^{N_\Gamma} (\ell_{n-} + \ell_{n+}) \psi_j^{\diamond}(\mathbf{p}_n)\right] . \quad (23)$$

The calculation of the desingularised value of the double layer fundamental solution normal derivative can be performed through the discretisation of the Equation (21) in the following simple way

$$\sum_{n=1}^{N_\Gamma} \frac{\ell_{n-} + \ell_{n+}}{2} \frac{\partial \psi_n^{\diamond}}{\partial \mathbf{n}_\Gamma}(\mathbf{p}_j) = 0 \,, \quad (24)$$

$$\frac{\partial \psi_j^{\diamond}}{\partial \mathbf{n}_\Gamma}(\mathbf{p}_j) = -\frac{1}{\ell_{j-} + \ell_{j+}}\left[\sum_{\substack{n=1 \\ n \neq j}}^{N_\Gamma} (\ell_{n-} + \ell_{n+}) \frac{\partial \psi_j^{\diamond}}{\partial \mathbf{n}_\Gamma}(\mathbf{p}_j)\right] . \quad (25)$$

The internal and external boundaries are given by a vector of points $\mathbf{p}_k; k = 1,2,...,N_\Gamma^B; B = I,E$. The length ℓ_k of the contour between the boundary points $\mathbf{p}_k$ and $\mathbf{p}_{k-1}$ is parametrised by the simple Euclidean distance

$$\ell_{k\pm} = \left[\left(p_{kx} - p_{k\pm 1x} \right)^2 + \left(p_{ky} - p_{k\pm 1y} \right)^2 \right]^{1/2}, \tag{26}$$

with the cyclic index conditions $k-1 = N_\Gamma^B;\ k=1,\ \ k+1=1;\ k = N_\Gamma^B$. The total Euclidean length ℓ_Γ of the boundary contour equals to

$$\ell_\Gamma = \sum_{k=1}^{N_\Gamma^B} \ell_k \ . \tag{27}$$

The derivatives of the fundamental solution can be calculated in the following indirect way. Let us assume a pure Dirichlet problem with all the boundary values set to a constant $\overline{\Phi}^D(\mathbf{p}) = c;\ \mathbf{p} \in \Gamma$. We obtain in this case

$$\Phi(\mathbf{p}_j) = c = \sum_{n=1}^{N_\Gamma} \psi_n(\mathbf{p}_j)\alpha_n^c, \tag{28}$$

$$\frac{\partial}{\partial p_\xi}\Phi(\mathbf{p}_j) = \sum_{n=1}^{N_\Gamma} \frac{\partial}{\partial p_\xi}\psi_n(\mathbf{p}_j)\alpha_n^c = 0;\ \xi = x,y\ . \tag{29}$$

The desingularised value of the partial derivative can be calculated as

$$\frac{\partial}{\partial p_\xi}\psi_j^\lozenge(\mathbf{p}_j) = -\frac{1}{\alpha_j^c}\sum_{\substack{n=1\\ n\neq j}}^{N_\Gamma} \frac{\partial}{\partial p_\xi}\psi_n^\lozenge(\mathbf{p}_j)\alpha_n^c; \xi = x,y\ . \tag{30}$$

The desingularised value of the normal derivative can be calculated from the desingularised values of the partial derivatives as

$$\frac{\partial \psi_j^\lozenge}{\partial \mathbf{n}_\Gamma}(\mathbf{p}_j) = \frac{\partial}{\partial p_x}\psi_j^\lozenge(\mathbf{p}_j)n_x(\mathbf{p}_j) + \frac{\partial}{\partial p_y}\psi_j^\lozenge(\mathbf{p}_j)n_y(\mathbf{p}_j). \tag{31}$$

This represents an alternative way how to arrive to Equation (25). The present approach does not require a direct integration, which distinguishes it from the approach with the single layer fundamental solution [9] where a direct BEM-like integration is needed.

4 Numerical example

Potential flow arround a circle is considered for a numerical example. The flow is confined to a square (exterior) region Γ^E, $p_x^- \le p_x \le p_x^+$, $p_y^- \le p_y \le p_y^+$ with $p_x^+ = -p_x^- = p_0$, $p_y^+ = -p_y^- = p_0$. The Dirichlet boundary conditions are defined at the square boundaries as

$$\overline{\Phi}^D(p_x, p_y) = v_0 p_y;\ p_x = p_x^\pm,\ p_y = p_y^\pm\ . \tag{32}$$

The potential field, defined from the boundary conditions (32) gives the following solution for the velocity field

$$v_{0x} = 0\ ,\quad v_{0y} = v_0\ . \tag{33,34}$$

A circular hole Γ^I (internal boundary) with the radius r_0, centered around point $\mathbf{p}_c$ with coordinates $p_{cx} = \left(p_x^+ + p_x^-\right)/2$, $p_{cy} = \left(p_y^+ + p_y^-\right)/2$ with the Neumann boundary conditions

$$\overline{\Phi}^N \left[p_x\left(\Gamma^I\right), p_y\left(\Gamma^I\right) \right] = 0; \; \mathbf{p} \in \Gamma^I \; . \tag{35}$$

is inserted into the square. The solution of the potential field for $r_0 \ll p_0$ equals

$$\Phi_{\text{ana}} = v_{0y}\left(p_y - p_{cy}\right)\left[1 + \frac{r_0^2}{\left(p_x - p_{cx}\right)^2 + \left(p_y - p_{cy}\right)^2}\right], \tag{36}$$

with v_{0y} defined from equation (46). The respective analytical solution for the velocity field is

$$v_{\text{ana}\,x} = -v_{0x}\frac{2 v_{0y} r_0^2 \left(p_x - p_{cx}\right)\left(p_y - p_{cy}\right)}{\left(p_x - p_{cx}\right)^2 + \left(p_y - p_{cy}\right)^2} \tag{37}$$

$$v_{\text{ana}\,y} = v_{0x}\left[1 + \frac{r_0^2}{\left(p_x - p_{cx}\right)^2 + \left(p_y - p_{cy}\right)^2}\right] - \frac{2 v_{0y} r_0^2 \left(p_y - p_{cy}\right)^2}{\left(p_x - p_{cx}\right)^2 + \left(p_y - p_{cy}\right)^2} \tag{38}$$

We set $r_0 = 0{,}1\,\text{m}$, $p_0 = 0{,}5\,\text{m}$ for geometry. The square sides are virtually divided into 50 equal length segments and the collocation points are put at each of the segment centres. The total number of discretisation points on the external boundary is set to $N^E = 200$. The circle is discretised by five different discretisations $N^I = 8,16,32,64,128$. The Root Mean Square (RMS) error of the MMFS solution is defined as

$$\Phi_{\text{rms}} = \sum_{n=1}^{N^I}\left\{\frac{1}{N^I}\left[\Phi(\mathbf{p}_n) - \Phi_{\text{ana}}(\mathbf{p}_n)\right]^2\right\}^{1/2}, \tag{39}$$

$$v_{\xi\,\text{rms}} = \sum_{n=1}^{N^I}\left\{\frac{1}{N^I}\left[v_\xi(\mathbf{p}_n) - v_{\xi\,\text{ana}}(\mathbf{p}_n)\right]^2\right\}^{1/2}; \xi = x, y, \tag{40}$$

$$v_{\text{rms}} = \left(v_{x\,\text{rms}}^2 + v_{y\,\text{rms}}^2\right)^{1/2}. \tag{41}$$

The RMS errors of the potential, velocity components, and absolute value of velocity are in case of MMFS given in Table 1 as a function of the discretisation density N^I. One can observe monotone convergence of the results with finer discretisation. A comparison with the results of the MFS (with artificial boundary), based on the single layer fundamental solution of the Laplace quation (the results are tabulated in [9]) shows, that the potential is in the case of MFS calculated more accurately as in the case of MMFS for $N^I = 8,16$. The velocity components and absolute value of velocity are better predicted by the present MMFS than in the classical MFS in all attempted discretisations.

5 Conclusions

In recent years, the MFS has proved to be an effective alternative to the boundary element methods in specific problems. Due to its advantages with respect to the simplicity of formulation and the fact that the distribution of the calculation nodes is truly meshless, the method is an ideal candidate for the moving and free boundary problems. Its main drawback represents the "artificial boundary issue". This issue has been in this work overcome through the MMFS concept, based on the double layer fundamental solution. The desingularisation of the potential has been in the present work made through the direct integral equation concept. The desingularisation of the spatial derivatives has been made in an indirect way through the constant potential field concept. Only the discretisation nodes on the real boundary are thus required. Both approaches differ from the previous two pioneering works on the subject [7,8]. In addition, this paper extends the MMFS to potential flow situations. The components of the flow field are calculated more precisely with the MMFS as with the MFS at all

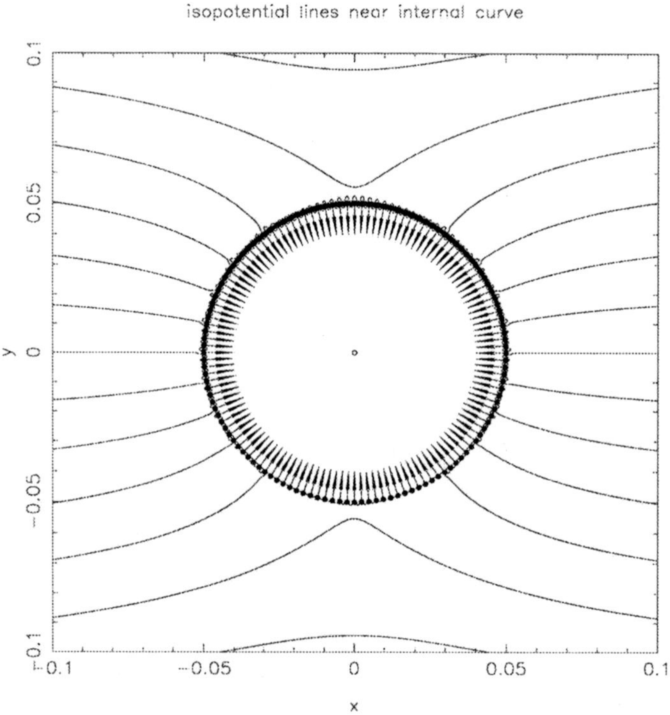

Figure 1: Calculated potential around the circle. The difference between the isopotential lines is 0,02 m^2/s.

discretisations used. The presented developments can be straightforwardly upgraded to axisymmetric problems [10] by inclusion of the axisymmetric fundamental solution. The flow physics can be extended to Navier-Stokes flow by the strategy, proposed in [11] which uses the dual reciprocity with radial basis functions. The axisymmetric radial basis functions, such as thin plate splines [12] and multiquadrics [13] can be used for this purpose in axisymmetry.

Table 1: Calculated error of the solution as a function of the number of collocation points on the circle boundary.

N^I	Φ_{rms}	v_{xrms}	v_{yrms}	v_{rms}
8	1.037098E-02	6.825820E-03	1.177813E-02	1.361309E-02
16	4.277747E-03	6.385956E-03	1.098282E-02	1.270444E-02
32	1.704762E-03	6.171696E-03	1.063205E-02	1.229351E-02
64	5.225166E-04	6.069899E-03	1.046481E-02	1.209776E-02
128	4.806703E-05	6.020260E-03	1.038311E-02	1.200219E-02

Acknowledgement

This paper forms a part of the project J2-0099: *Multiscale Modelling and Simulation of Liquid-Solid Systems*, sponsored by the Slovenian Research Agency. Financial support is kindly acknowledged.

References

[1] Wrobel LC, Aliabadi MH. *The Boundary Element Method*, J. Wiley, Chichester, 2002.

[2] Fairweather G, Karageorghis A. The method of fundamental solutions for elliptic boundary value problems, *Advances in Computational Mathematics*, 1998; **9**: 69–95.

[3] Golberg MA, Chen CS. *Discrete Projection Methods for Integral Equations*, CMP, Southampton, 1997.

[4] Golberg MA, Chen CS. The method of fundamental solutions for potential, Helmholtz and diffusion problems, *Boundary Integral Methods - Numerical and Mathematical Aspects*, ed. Golberg, M. A., CMP, Southampton, 1998: 103–176.

[5] Johnston RL, Fairweather G. The method of fundamental solutions for problems in potential flow, *Applied Mathematical Modelling*, 1984; **8**: 265–270.

[6] Šarler B. Solution of a two-dimensional bubble shape in potential flow by the method of fundamental solutions, *Engineering Analysis with Boundary Elements*, 2006; **30**: 227–235.

[7] Young DL, Chen KH, Lee CW. Novel meshless method for solving the potential problems with arbitrary domain, *Journal of Computational Physics*, 2005; **209**: 290–322.

[8] Young DL, Chen JT, Kao JH. A modified method of fundamental solutions with source on the boundary for solving Laplace equations with circular and arbitrary domains, *CMES: Computer Modelling in Engineering and Sciences*, 2007; **19**: 197–221.

[9] Šarler B. Chapter 15: Modified method of fundamental solutions for potential flow problems. Chen CS, Karageorghis A, Smyrlis YS (Eds.), *Method of Fundamental Solutions*, Tech Science Press, Encinio, 2008; (to appear).

[10] Karageorghis A, Fairweather G. The method of fundamental solutions for axisymmetric potential problems, *International Journal of Numerical Methods in Engineering*, 1999; **44**: 1653–1669.

[11] Šarler B. Towards a mesh-free solution of transport phenomena, *Engineering Analysis with Boundary Elements*, 2002; **26**: 731–738.

[12] Šarler B. Axisymmetric augmented thin plate splines, *Engineering Analysis with Boundary Elements*, 1998; **21**: 81–85.

[13] Šarler B, Jelić N, Kovačević I, Lakner M, Perko J. Axisymmetric multiquadrics, *Engineering Analysis with Boundary Elements*, 2006; **30**: 137–142.

On the use of integrated RBFs in Galerkin approximation for elliptic problems

N. Mai-Duy & T. Tran-Cong
Computational Engineering and Science Research Centre,
University of Southern Queensland, Toowoomba, QLD 4350, Australia

Abstract

This paper presents a new radial-basis-function (RBF) technique for solving elliptic differential equations (DEs). Integrated RBF networks (IRBFN) are constructed to satisfy (a) the boundary conditions in a local sense using the point-collocation formulation, and (b) the DE in a global sense using the Galerkin formulation. Several numerical examples are presented to demonstrate the attractiveness of the present implementation.
Keywords: integrated RBFs, Galerkin formulation, Neumann boundary conditions, multiple boundary conditions, domain decomposition.

1 Introduction

Many discretisation techniques can be viewed as variants of the method of weighted residuals that can be stated in three well-known formulations, namely the strong, weak and inverse statements [1]. By means of weighting functions in a statement, the residuals for the DE and boundary conditions are made small in some senses. Two popular ways used are (i) the point-collocation approach, where the residuals are zero at certain points and (ii) the Galerkin-type approach, in which the residuals are zero in the mean. Each approach has some advantages in certain areas of application. The former is cost-effective as no integrations are required, while the latter has a smoothing capability owing to its integral nature.

RBF collocation methods are considered as a powerful tool for the interpolation of scattered data as well as for the solution of differential problems [2]. The methods are capable of approximating arbitrarily-well continuous functions. A number of RBFs such as multiquadric and Gaussian functions have spectral approximation power. However, the condition number of the RBF interpolation matrix also grows

WIT Transactions on Modelling and Simulation, Vol 47, © 2008 WIT Press
www.witpress.com, ISSN 1743-355X (on-line)
doi:10.2495/BE080171

rapidly with respect to (a) the decrease in distance between the RBF centres and (b) the increase in the RBF width. RBF collocation methods thus, in practice, suffer from a trade-off between accuracy and stability. Moreover, there is a gap in accuracy between the RBF solutions to Neumann- and Dirichlet-type boundary-value problems. To improve the numerical stability of a RBF solution, there have been a number of schemes proposed in the literature: for example, (a) preconditioning the system matrix; (b) constructing the approximations locally; (c) using compactly-supported RBFs; and (d) employment of domain decompositions. Recently, an approximation scheme, which is based on point collocation, Cartesian grids and one-dimensional IRBFNs, has been proposed in [3, 4]. A problem domain, which can be regular or irregular, is discretised by a Cartesian grid. Along grid lines, 1D-IRBFNs are constructed to satisfy the governing DE together with boundary conditions in an exact manner. This scheme allows a larger number of nodes to be employed.

There are very few papers on the use of RBFs in the context of Galerkin approximation [2]. In this study, we present a new numerical scheme, which is based on the Galerkin formulation and 1D-IRBFNs, for solving elliptic problems. From a Galerkin-approach point of view, it will be shown that the present derivative boundary conditions are imposed in an exact manner. From a RBF-approach point of view, it will be shown that (a) the proposed method is able to yield almost the same levels of accuracy for the solutions of Neumann- and Dirichlet-type problems and that (b) its accuracy is considerably better that of the 1D-IRBFN collocation method. An additional attractiveness of the proposed technique is that it facilitates a higher-order continuity of the solution across the subdomain interfaces.

The paper is organised as follows. Brief reviews of the Galerkin formulation and 1D-IRBFNs are given in Sections 2 and 3, respectively. The Galerkin 1D-IRBFN method is presented in Section 4, followed by several numerical examples in Section 5. Section 6 concludes the paper.

2 Galerkin approach

The Galerkin-type approach is well documented in the literature (e.g. [1]). A brief review of this approach is given below.

Consider a boundary-value problem defined by a linear DE and its boundary conditions

$$L(\bar{u}) = 0, \quad \mathbf{x} \in \Omega, \quad B(\bar{u}) = 0, \quad \mathbf{x} \in \Gamma, \tag{1}$$

where $\bar{u}$ is the field/dependent variable (the overbar denotes the exact solution), L and B prescribed known operators, Ω the domain of interest and Γ the boundaries of the domain Ω.

An approximate solution, denoted by u, to (1) can be sought in the form

$$\bar{u}(\mathbf{x}) \approx u(\mathbf{x}) = \sum_{i=1}^{N} \alpha_i \phi_i(\mathbf{x}), \tag{2}$$

where $\{\alpha_i\}_{i=1}^N$ is the set of unknown coefficients and $\{\phi_i(\mathbf{x})\}_{i=1}^N$ the set of linearly-independent functions. The terms ϕ_i are usually referred to as the trial/basis/approximating functions. Assume that a function u is constructed to satisfy the DE at every point on the domain Ω, it leads to

$$\int_\Omega wL(u)d\Omega = 0, \tag{3}$$

for any function w that is bounded on Ω. Similarly, assume that the approximate solution u also satisfies the boundary conditions, it follows that

$$\int_\Gamma \tilde{w}B(u)d\Gamma = 0, \tag{4}$$

for any bounded function $\tilde{w}$. The functions w and $\tilde{w}$ are often referred to as the weighting/test functions. It can be seen that the system defined by (1) is equivalent to the following integral statement

$$\int_\Omega wL(u)d\Omega + \int_\Gamma \tilde{w}B(u)d\Gamma = 0, \tag{5}$$

that is satisfied for all bounded functions w and $\tilde{w}$.

If the weighting functions w and $\tilde{w}$ have sufficient degrees of continuity, integrations by parts can be applied to derivative terms in (5), leading to other integral statements, namely the weak and inverse forms, that can be expressed as

$$\int_\Omega C(w)D(u)d\Omega + \int_\Gamma E(\tilde{w})F(u)d\Gamma = 0, \tag{6}$$

where the order of continuity required for the u solution is reduced. One can thus use either (5) or (6) to determine the approximate solution u. For the Galerkin-type approach, the weighting functions are chosen from the same set of functions as the trial functions.

3 One-dimensional integrated RBFNs

Consider a univariate function $f(x)$. The basic idea of the integral RBF scheme [3, 4] is to decompose a pth-order derivative of the function f into RBFs

$$\frac{d^p f(x)}{dx^p} = \sum_{i=1}^N w_i g_i(x) = \sum_{i=1}^N w_i I_i^{(p)}(x), \tag{7}$$

where $\{w_i\}_{i=1}^N$ is the set of network weights, and $\{g_i(x)\}_{i=1}^N \equiv \{I_i^{(p)}(x)\}_{i=1}^N$ the set of RBFs. Lower-order derivatives and the function itself are then obtained through integration

$$\frac{d^{p-1} f(x)}{dx^{p-1}} = \sum_{i=1}^N w_i I_i^{(p-1)}(x) + c_1, \ldots, \tag{8}$$

$$f(x) = \sum_{i=1}^{N} w_i I_i^{(0)}(x) + c_1 \frac{x^{p-1}}{(p-1)!} + c_2 \frac{x^{p-2}}{(p-2)!} + \cdots + c_{p-1}x + c_p, \quad (9)$$

where $I_i^{(p-1)}(x) = \int I_i^{(p)}(x)dx, \ldots, I_i^{(0)}(x) = \int I_i^{(1)}(x)dx$, and $\{c_1, c_2, \ldots, c_p\}$ are the constants of integration.

Unlike conventional differential schemes, the starting point of the integral scheme can vary in use, depending on the particular application under consideration. The scheme is said to be of order p, denoted by IRBFN-p, if the pth-order derivative is taken as the starting point.

Evaluation of (7)–(9) at a set of collocation points $\{x_j\}_{j=1}^N$ leads to

$$\widehat{\frac{d^p f}{dx^p}} = \widehat{\mathcal{I}}_{[p]}^{(p)}\widehat{\alpha}, \ldots, \widehat{f} = \widehat{\mathcal{I}}_{[p]}^{(0)}\widehat{\alpha},$$

where the subscript [.] and superscript (.) are used to denote the order of IRBFN and the order of the corresponding derivative function, respectively;

$$\widehat{\mathcal{I}}_{[p]}^{(p)} = \begin{bmatrix} I_1^{(p)}(x_1), & \ldots, & I_N^{(p)}(x_1), & 0, & 0, & \ldots, & 0, & 0 \\ I_1^{(p)}(x_2),, & \ldots, & I_N^{(p)}(x_2), & 0, & 0, & \ldots, & 0, & 0 \\ \ldots & \ldots & \ldots & \ldots & \ldots & \ldots & \ldots & \ldots \\ I_1^{(p)}(x_N), & \ldots, & I_N^{(p)}(x_N), & 0, & 0, & \cdots, & 0, & 0 \end{bmatrix}, \ldots$$

$$\widehat{\mathcal{I}}_{[p]}^{(0)} = \begin{bmatrix} I_1^{(0)}(x_1), & \ldots, & I_N^{(0)}(x_1), & \frac{x_1^{p-1}}{(p-1)!}, & \frac{x_1^{p-2}}{(p-2)!}, & \ldots, & x_1, & 1 \\ I_1^{(0)}(x_2), & \ldots, & I_N^{(0)}(x_2), & \frac{x_2^{p-1}}{(p-1)!}, & \frac{x_2^{p-2}}{(p-2)!}, & \ldots, & x_2, & 1 \\ \ldots & \ldots & \ldots & \ldots & \ldots & \ldots & \ldots & \ldots \\ I_1^{(0)}(x_N), & \ldots, & I_N^{(0)}(x_N), & \frac{x_N^{p-1}}{(p-1)!}, & \frac{x_N^{p-2}}{(p-2)!}, & \ldots, & x_N, & 1 \end{bmatrix};$$

$$\widehat{\alpha} = (w_1, w_2, \ldots, w_N, c_1, c_2, \ldots, c_p)^T;$$

and

$$\widehat{\frac{d^k f}{dx^k}} = \left(\frac{d^k f_1}{dx^k}, \frac{d^k f_2}{dx^k}, \ldots, \frac{d^k f_N}{dx^k}\right)^T, \widehat{f} = (f_1, f_2, \ldots, f_N)^T,$$

in which $d^k f_j/dx^k = d^k f(x_j)/dx^k$ and $f_j = f(x_j)$ with $k = \{1, 2, \ldots, p\}$ and $j = \{1, 2, \ldots, N\}$.

4 Galerkin IRBFN technique

For Galerkin finite-element techniques, a weak statement (6), where the continuity requirement for the field variable u is reduced, is a preferred option. Piecewise polynomials of low order such as linear and quadratic interpolations are generally used as approximating and weighting functions in numerous small subdomains called elements. In the case that the shape functions φ_i are algebraic polynomials, only the field variable changes continuously throughout the entire domain, and high derivatives (e.g. second derivatives for linear elements) are not defined. Essential boundary conditions are incorporated into the approximate solution prior to the process of discretising the DE, while natural boundary conditions are imposed by means of weighted residual (i.e. the second term in (6)). It should be emphasised that the natural boundary conditions in the weak formulation are approximated rather than identically satisfied. In engineering practice, such a partial satisfaction of the boundary conditions tends to give poor results for surface fluxes or tractions which make the overall results unreliable for many cases [1].

In the present Galerkin 1D-IRBFN technique, we use a Cartesian grid to generate the finite trial and test spaces. One dimensional IRBFNs are employed to represent the field variable and its derivatives on grid lines. The RBF solutions are constructed to satisfy the boundary conditions using the point-collocation approximation and the governing DE using the Galerkin approximation. A distinguishing feature here is that the networks are sought to satisfy a priori the derivative boundary conditions in an exact manner. There is thus no need to use the second term in (5) and (6). As the trial functions are infinitely-differentiable global functions, the present Galerkin 1D-IRBFN technique permits the employment of (5) to solve the differential problem of any order. Moreover, any derivative of the field variable is defined and continuous throughout the entire domain.

From an engineering viewpoint, one would prefer to work in the physical space. The present approximate solution is sought in terms of nodal variable values rather than the usual network weights. The boundary conditions including derivative information are imposed through the conversion process of the network-weight space into the physical space. RBFNs involve two types of data sets, namely centre and collocation points. In the context of point-collocation approximation, RBFNs tend to result in the most accurate approximations when the two sets of points are identical. Here, the collocation points are chosen to be the centres themselves. Unlike conventional differential formulations, the integral RBF formulation has the ability to generate additional coefficients (the constants of integration). This feature thus facilitates the addition of extra equations to the conversion system to represent extra information such as the natural boundary conditions and even the governing equation at the boundary points. The presence of integration constants thus guarantees that all RBFs are used for function approximation. In contrast, for conventional differentiated RBFNs, the enforcement of derivative function values is done at the price of the non-consideration of the function at some RBF centres, which significantly deteriorates the accuracy of the RBF scheme.

Consider a grid line. The conversion system for an 1D-IRBFN scheme of order p can be described as

$$\begin{pmatrix} \widehat{u} \\ \widehat{e} \end{pmatrix} = \begin{bmatrix} \widehat{\mathcal{I}}^{(0)}_{[p]} \\ \widehat{\mathcal{K}} \end{bmatrix} \widehat{\alpha} = \mathcal{C}\widehat{\alpha}, \tag{10}$$

where $\widehat{e}$, whose length can be up to p, is a vector representing extra information (e.g. normal derivative boundary conditions); $\widehat{e} = \widehat{\mathcal{K}\alpha}$; $\widehat{u}$ and $\widehat{\mathcal{I}}^{(0)}_{[p]}$ defined as before; and $\mathcal{C}$ the conversion matrix. It can be seen from (10) that the approximate solution u is collocated at the whole set of centres. Solving (10) for $\widehat{\alpha}$ yields

$$\widehat{\alpha} = \mathcal{C}^{-1} \begin{pmatrix} \widehat{u} \\ \widehat{e} \end{pmatrix}, \tag{11}$$

where $\mathcal{C}^{-1}$ is the inverse or pseudo-inverse of $\mathcal{C}$, depending on its dimension. Substitution of (11) into (7)–(9) leads to

$$u(x) = (I_1^{(0)}(x), I_2^{(0)}(x), \ldots)\mathcal{C}^{-1} \begin{pmatrix} \widehat{u} \\ \widehat{e} \end{pmatrix}, \tag{12}$$

$$\cdots\cdots\cdots\cdots\cdots\cdots$$

$$\frac{\partial^p u(x)}{\partial x^p} = (I_1^{(p)}(x), I_2^{(p)}(x), \ldots)\mathcal{C}^{-1} \begin{pmatrix} \widehat{u} \\ \widehat{e} \end{pmatrix}. \tag{13}$$

They can be rewritten in the form

$$u(x) = \sum_{i=1}^{N} \varphi_i(x) u_i + \varphi_{N+1}(x) e_1 + \varphi_{N+2}(x) e_2 + \cdots, \tag{14}$$

$$\cdots\cdots\cdots\cdots\cdots\cdots$$

$$\frac{\partial^p u(x)}{\partial x^p} = \sum_{i=1}^{N} \frac{d^p \varphi_i(x)}{dx^p} u_i + \frac{d^p \varphi_{N+1}(x)}{dx^p} e_1 + \frac{d^p \varphi_{N+2}(x)}{dx^p} e_2 + \cdots. \tag{15}$$

The Galerkin weighting process applied to (1) produces the results

$$\int_\Omega \varphi_i L(u) = 0, \tag{16}$$

where the values of i depend on the problem under consideration as will be discussed later. The system of equations, (16), can then be used to solve for the nodal value of the variable u.

5 Numerical results

Due to space limitation, only numerical results for second- and fourth-order ODEs are presented here. An ODE of order p is discretised by the 1D-IRBFN-p scheme using multiquadrics (MQ). To generate finite spaces for the trial and test functions, uniformly-distributed MQ centres are employed. The MQ width is simply chosen to be the centre spacing. A 1D-IRBFN-based collocation method is also employed to provide a useful basis for comparison. The accuracy of an approximate solution is measured by means of the discrete relative L_2 norm, denoted by N_e.

5.1 Second-order ODE

Find an approximate solution to the ODE

$$\frac{d^2\bar{u}}{dx^2} + \bar{u} + x = 0, \tag{17}$$

on the interval $0 \leq x \leq 1$. The exact solution of (17) is assumed to be $\bar{u} = \sin(x)/\sin(1) - x$, from which one can easily derive the boundary values at $x = 0$ and $x = 1$. We employ 45 sets of uniformly-distributed centres, varying from 3 to 91 with increment of 2. Two types of boundary conditions are considered.

Dirichlet boundary conditions: Both $\mathcal{K}$ and $\widehat{e}$ are set to null. An approximate solution simply takes the form $u(x) = \sum_{i=1}^{N} u_i \varphi_i(x)$. The weighting functions are chosen to be the trial functions that are associated with the unknown nodal values of u. For this case, they are $\{\varphi_2(x), \ldots, \varphi_{N-1}(x)\}$. Equation (16) leads to a determinate symmetric system of equations A for $(N-2)$ unknowns (i.e. the values of u at the interior points). It can be seen from Figure 1 (left) that the Galerkin technique yields much more accurate results and converges faster than the collocation technique.

Dirichlet ($x = 0$) and Neumann ($x = 1$) boundary conditions: We employ one extra equation to represent the derivative value at $x = 1$. Expression (14) becomes $u(x) = \sum_{i=1}^{N} u_i \varphi_i(x) + (du_N/dx)\varphi_{N+1}(x)$ that also contains derivative information. The approximate solution thus satisfy a priori both the Dirichlet and Neumann boundary conditions in an exact manner. The unknown vector consists of the values of u at $\{x_j\}_{j=2}^{N}$. As a result, $\{\varphi_2(x), \ldots, \varphi_N(x)\}$ are taken as the weighting functions. Again, the Galerkin approach outperforms the collocation approach regarding accuracy (Figure 1 (right)).

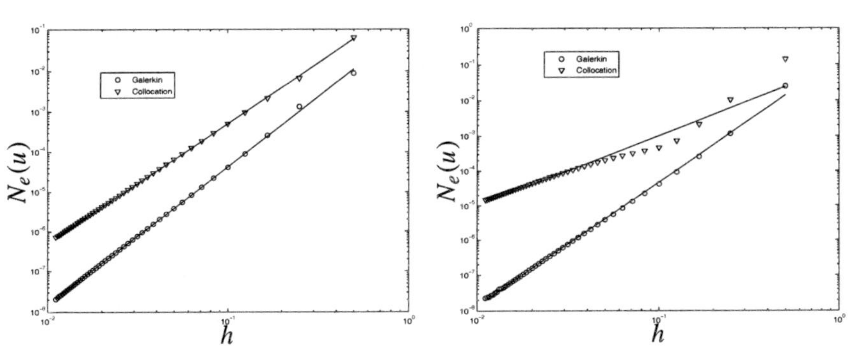

Figure 1: Second-order ODE: Error $N_e(u)$ versus the centre spacing h for the Galerkin and collocation solutions. For Dirichlet conditions (left), they converge as $O(h^{3.46})$ and $O(h^{2.96})$, respectively. For Neumann conditions (right), they are $O(h^{3.54})$ and $O(h^{1.98})$.

WIT Transactions on Modelling and Simulation, Vol 47, © 2008 WIT Press
www.witpress.com, ISSN 1743-355X (on-line)

It has been generally observed that the RBF results for the case of Neumann boundary conditions are generally much less accurate than for the case of Dirichlet boundary conditions. An attractive point here is that the present RBF method yields essentially the same degrees of accuracy for both types of boundary conditions (Figure 1). For both cases, values of the matrix condition number are relatively low, varying from $O(10^1)$ to $O(10^4)$.

5.2 Fourth-order ODE

This example is governed by the biharmonic equation

$$\frac{d^4\bar{u}}{dx^4} + \frac{d^2\bar{u}}{dx^2} + (4k^3 - 2k)\cos(kx) - x(k^4 - k^2)\sin(kx) = 0, \tag{18}$$

on the interval $-1/2 \leq x \leq +1/2$, and Dirichlet boundary conditions. The exact solution to this problem is chosen to be $\bar{u} = x\sin(kx)$. We employ $k = 7\pi/2$ that makes all boundary data nonzero. The present conversion process involves the enforcement of u at the whole set of centres and du/dx at the two boundary points, from which the approximate solution will take the form $u(x) = \sum_{i=1}^{N} u_i\varphi_i(x) + (du_1/dx)\,\varphi_{N+1}(x) + (du_N/dx)\,\varphi_{N+2}(x)$. The system matrix for solving $\{u_i\}_{i=2}^{N-1}$ is then generated using the weighting functions $\{\varphi_2(x), \ldots, \varphi_{N-1}(x)\}$. A number of uniform centre sets, namely $5, 7, 9, \ldots, 45$ points, are employed. Figure 2 shows that the accuracy of the Galerkin method is far superior to that of the collocation method. The present Galerkin matrices have the condition number of $O(10^1)$ to $O(10^5)$.

5.3 Domain decomposition

The most time-consuming part of the proposed technique lies in the process of computing volume integrals to form the algebraic system. One strategy to overcome this problem is to use domain decomposition. A substructuring technique

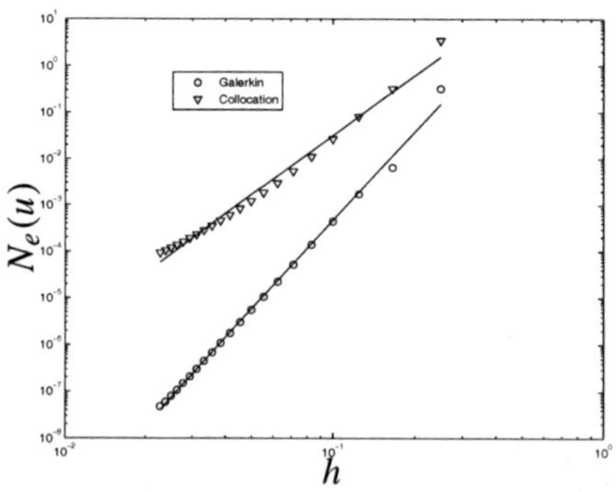

Figure 2: Fourth-order ODE: $N_e(u)$ versus h for the Galerkin and collocation solutions. They converge as $O(h^{6.31})$ and $O(h^{4.24})$, respectively.

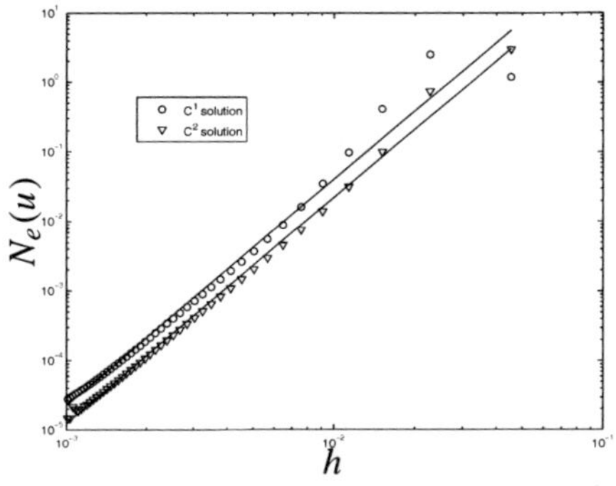

Figure 3: Domain decomposition, 11 subdomains: Errors of the present C^1 and C^2 Galerkin solutions.

will be applied here. The interface system can be constructed by requiring continuity of the field variable and its derivatives of order up to $(p-1)$ (p—the order of the DE) across the subdomain interfaces.

The combination of the substructuring technique and the proposed method that is presented above will lead to an approximate solution u that is a C^1 function for second-order problems and C^3 for fourth-order problems.

However, as shown in [5], a C^p solution can be achieved if the DE is enforced to be satisfied at the interface points in the subdomain solutions. This satisfaction can be made through the transformation of the network-weight space into the physical space.

Two versions of the present multidomain Galerkin technique are applied to the following ODE

$$\frac{d^2\bar{u}}{dx^2} + \frac{d\bar{u}}{dx} + \bar{u} = -\exp(-5x)[9979\sin(100x) + 900\cos(100x)], \quad 0 \leq x \leq 1, \tag{19}$$

with Dirichlet boundary conditions. The exact solution is chosen to be $\bar{u}(x) = \sin(100x)\exp(-5x)$, which is highly oscillatory. The domain is partitioned into 11 subdomains that are then identically represented using grids of $\{3, 5, \ldots, 91\}$ uniform points. Figure 3 clearly shows that the present C^2 solution is, as expected, more stable and accurate than the C^1 solution.

6 Concluding remarks

In this paper, a numerical technique, based on 1D-IRBFNs and Galerkin approximation, is developed for solving elliptic differential equations. In contrast to Galerkin finite-element techniques, the Neumann boundary conditions are presently imposed in an exact manner. Unlike conventional RBF techniques, the RBF approximations are presently constructed locally on grid lines through integration and they are expressed in terms of nodal variable values. The proposed technique produces a system of algebraic equations that is often symmetric and has a relatively-low matrix condition number. To avoid the problem of high cost associated with the evaluation of volume integrals, the use of domain decomposition is discussed, where continuity order can be improved. Numerical results have shown that (a) the present technique achieves a high rate of convergence, (b) its accuracy is much higher than that of the integrated-RBF collocation technique, and (c) the obtained solutions have similar levels of accuracy for both Dirichlet- and Neumann-type boundary-value problems.

Acknowledgement

This work is supported by the Australian Research Council.

References

[1] Brebbia, C.A. & Dominguez, J., *Boundary Elements–An Introductory Course*, Computational Mechanics Publications: Southampton, 1992.

[2] Fasshauer G.E., *Meshfree Approximation Methods With Matlab*, World Scientific Publishers: Singapore, 2007.

[3] Mai-Duy, N. & Tanner R.I., A collocation method based on one-dimensional RBF interpolation scheme for solving PDEs. *Int. J. of Numerical Methods for Heat & Fluid Flow*, **17(2)**, pp. 165–186, 2007.

[4] Mai-Duy N. & Tran-Cong T., A Cartesian-grid collocation method based on radial-basis-function networks for solving PDEs in irregular domains. *Numerical Methods for Partial Differential Equations*, **23(5)**, pp. 1192–1210, 2007.

[5] Mai-Duy N. & Tran-Cong T., A multidomain integrated radial basis function collocation method for elliptic problems. *Numerical Methods for Partial Differential Equations* (in press).

Boundary element dual reciprocity method with overlapping sub-domains

B. Thanh Tu & V. Popov
Wessex Institute of Technology, Environmental Fluid Mechanics, Ashurst Lodge, Southampton, UK

Abstract

In this paper a numerical approach based on the Boundary Element Method (BEM) with overlapping sub-domains is presented. The approach simplifies the assembly of the equations arising from the BEM sub-domain methods and at the same time reduces the problems with near-singular or singular integrals that may arise in the integral equations. The overlapping numerical approach is tested on a convection-diffusion problem. The approach is implemented in combination with the Dual Reciprocity Method (DRM) with two different Radial Basis Functions (RBFs) and the results are compared with the previous results obtained using the DRM-MD approach, showing good accuracy and convergence.

Keywords: boundary element method, domain decomposition, overlapping sub-domains.

1 Introduction

The present formulation employs domain decomposition, but unlike the previous BEM sub-domain methods it applies an overlapping approach which eliminates the Interface Boundary Conditions (IBCs), simplifies assembly of the equations and reduces the problems arising from near-singular or singular integrals. It also leads to a closed system of equations while the BEM sub-domain techniques with continuous elements lead to over-determined systems.

When using domain decomposition in three dimensions (3D) and for the case where the number of sub-domains is large, imposing the IBCs may be complicated since several sub-domains would generally share the same node (vertex). In the BEM the IBCs include the flux conservation, which increases the

WIT Transactions on Modelling and Simulation, Vol 47, © 2008 WIT Press
www.witpress.com, ISSN 1743-355X (on-line)
doi:10.2495/BE080181

complexity and therefore sometimes discontinuous elements are used in order to simplify the implementation of IBCs.

Another problem that often arises is the distortion in the geometry of some of the sub-domains due to complexity of the mesh. This may produce worsening of the condition number of the system of equations because of the proximity of the nodes and could introduce singularity or near-singularity in the integrals due to the distorted geometry of some of the sub-domains.

The proposed approach is quite general and can be applied to various BEM formulations, e.g., domain integration, DRM, multiple reciprocity method, etc. It eliminates the above mentioned problems due to IBCs and distorted sub-domains in the mesh.

2 The boundary element dual reciprocity method

In this section the boundary element dual reciprocity method is introduced though the approach will be effective with other BEM formulations as well. Let us consider the following equation:

$$\nabla u^2(r) = b\left(r, u(r), \frac{\partial u(r)}{\partial x_i}, \frac{\partial u(r)}{\partial t} \right) \tag{1}$$

where $u(r)$ is a potential field, r is a position vector, x_i is component of r, and t is time. Given a point r inside Ω, by applying the Green formula to equation (1) in Ω bounded by Γ, (1) can be transformed into the following integral form:

$$\lambda(r)u(r) + \int_{\Gamma} q^*(r,\xi)u(\xi)d\Gamma_\xi - \int_{\Gamma} u^*(r,\xi)q(\xi)d\Gamma_\xi = \int_{\Omega} u^*(r,\xi)b(\xi)d\Omega_\xi \tag{2}$$

Here $u^*(r,\xi)$ is the fundamental solution of the Laplace problem which is given by

$$u^*(r,\xi) = \frac{1}{2\pi} ln\left(\frac{1}{R}\right) \tag{3}$$

for a 2D problems and

$$u^*(r,\xi) = \frac{1}{4\pi}\frac{1}{R} \tag{4}$$

for 3D problems, where R is the distance from the point of application of the concentrated unit source to any other point under consideration, i.e. $R = |r - \xi|$ and $q(\xi) = \partial u(\xi)/\partial n$ and $q^*(r,\xi) = \partial u^*(r,\xi)/\partial n$. The constant $\lambda(r)$ has value from 0 to 1 being equal to 1/2 for smooth boundaries and 1 if the source point r is inside the domain.

By applying the DRM approach as described in [1], one finally arrives at a boundary only integral representation formula

$$\begin{aligned} &\lambda(r)u(r) + \int_{\Gamma} q^*(r,\xi)u(\xi)d\Gamma_\xi - \int_{\Gamma} u^*(r,\xi)q(\xi)d\Gamma_\xi = \\ &\sum_{k=1}^{J_{bn}+J_{in}} \left\{ \alpha_k \left(\lambda(r)\hat{u}(r,\eta_k) + \int_{\Gamma} q^*(r,\xi)\hat{u}(\xi,\eta_k)d\Gamma_\xi - \int_{\Gamma} u^*(r,\xi)\hat{q}(\xi,\eta_k)d\Gamma_\xi \right) \right\} \end{aligned} \tag{5}$$

The DRM approximation employs J_{bn} nodes on the boundary and J_{in} nodes inside the domain.

2.1 Treatment of partial derivatives in the DRM

In the classical DRM approach, in order to obtain expression for the partial derivatives, the potential at a point r is expressed in terms of an approximation function f, in a similar way as it is done for b [1]

$$u(\xi) \cong \sum_{k=1}^{J_{bn}+J_{in}} f(\xi,\eta_k)\beta_k \tag{6}$$

or expressed in matrix form:

$$\mathbf{u} = \mathbf{F}\boldsymbol{\beta} \tag{7}$$

Differentiating both sides in (7) yields:

$$\frac{\partial \mathbf{u}}{\partial x_i} = \frac{\partial \mathbf{F}}{\partial x_i}\boldsymbol{\beta} \tag{8}$$

From equation (7) the following expression can be derived $\boldsymbol{\beta} = \mathbf{F}^{-1}\mathbf{u}$ which can be substituted into (8) producing:

$$\frac{\partial \mathbf{u}}{\partial x_i} = \frac{\partial \mathbf{F}}{\partial x_i}\mathbf{F}^{-1}\mathbf{u} \tag{9}$$

Recently Natalini and Popov [2] proposed another approach which has been implemented for 3D problems as well, which uses the normal derivative before the node q_i^b, and the normal derivative after the node q_i^a to obtain the partial derivatives in the DRM-MD approach.

3 Non-overlapping (classical) BEM sub-domain technique

In this section the basic principles of the non-overlapping BEM sub-domain technique are introduced. In Figure 1 a case is presented where two sub-domains, 1 and 2, share a common interface where interface boundary/matching conditions are required in order to close the system of equations arising from the BEM sub-domain approach. The interface boundary conditions (IBCs) for a potential problem establish the continuity of the potential at every node at the interface, in this case node I, as shown in (10)

$$u_1(r_I) = u_2(r_I) \tag{10}$$

and the conservation of flux, as shown in (11)

$$\Phi_{12}(u_1,q_{12}) = -\Phi_{21}(u_2,q_{21}) \tag{11}$$

where the expression for the flux Φ in (11) depends on the physical problem under consideration. The situation with node J is more complicated, as this node is shared between four sub-domains. For homogeneous domains and when continuous boundary elements are used the IBCs yield to an over-determined system, which can be avoided by using discontinuous elements on the expense of increased number of degrees of freedom.

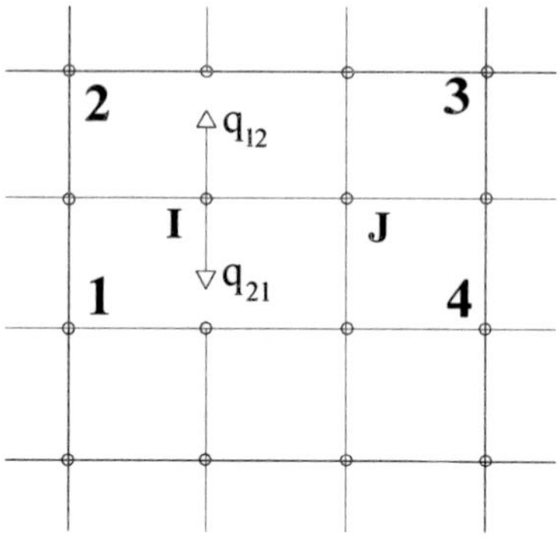

Figure 1: Example of sub-domains in a mesh.

4 Overlapping BEM sub-domain technique

The overlapping BEM sub-domain technique places the source point in the interior of the sub-domain. The main idea is to avoid placing the source on the boundary in order to avoid near-singular and/or singular integrals. In Figure 2 one can see two sub-domains, the first one being K-O-P-L with a source point in M and the second one being I-M-N-J with a source point in L, where the overlapping part of the two sub-domains is marked between points M and L.

The proposed formulation solves in each interior node three integral equations in order to obtain the potential u, and the partial derivatives $\partial u/\partial x_j$. The equations for derivatives $\partial u/\partial x_j$ are obtained by differentiating equation (5) in respect to x_j, where x_j are components of r.

The expression for partial derivatives is shown in (12), where the integrals are evaluated over the boundary Γ_i of the i-th sub-domain:

$$\frac{\partial u(r)}{\partial x_j} = -\int_{\Gamma_i} \frac{\partial q^*(r,\xi)}{\partial x_j} u(\xi) d\Gamma_\xi + \int_{\Gamma_i} \frac{\partial u^*(r,\xi)}{\partial x_j} q(\xi) d\Gamma_\xi + \\ \sum_{k=1}^{J_{bn}+J_{in}} \left\{ \alpha_k \left(\frac{\partial \hat{u}(r,\eta_k)}{\partial x_j} + \int_{\Gamma_i} \frac{\partial q^*(r,\xi)}{\partial x_j} \hat{u}(\xi,\eta_k) d\Gamma_\xi - \int_{\Gamma_i} \frac{\partial u^*(r,\xi)}{\partial x_j} \hat{q}(\xi,\eta_k) d\Gamma_\xi \right) \right\} \tag{12}$$

The discretized form of equation (12) for the unknown $\partial u/\partial x_j$ at node i is given as:

$$\frac{\partial u_i}{\partial x_j} = \frac{\partial u_i}{\partial x_j} = -\sum_{k=1}^{J_{bn}} H_{ik,j} u_k + \sum_{k=1}^{J_{bn}} G_{ik,j} q_k + \sum_{k=1}^{J_{bn}+J_{in}} \left\{ \alpha_k \left(\sum_{m=1}^{J_{bn}} H_{im,j} \hat{u}_{mk} - \sum_{m=1}^{J_{bn}} G_{im,j} \hat{q}_{mk} + c_i \hat{u}_{ik} \right) \right\} \tag{13}$$

where $H_{ik,j}$ and $G_{ik,j}$ are the influence coefficients obtained from the integrations of $(\partial q^*/\partial x_j)$ and $(\partial u^*/\partial x_j)$, respectively, at each boundary element. The normal derivative q in the equation (13) can be written as:

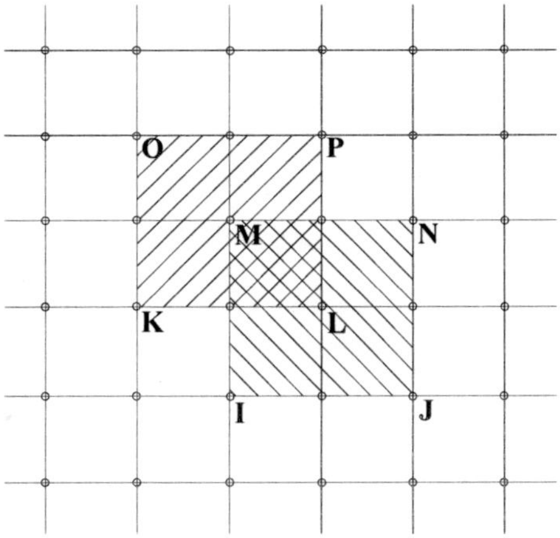

Figure 2: Part of the mesh used in the overlapping BEM sub-domain technique.

$$q = \frac{\partial u}{\partial n} = \nabla u \cdot \vec{n} = \sum_k \frac{\partial u}{\partial x_k} n_k \tag{14}$$

where n_k are components of the unit normal vector. According to the above (5) can be written as:

$$u_i + \sum_{j=1}^{J_{bn}} H_{ij} u_j - \sum_m \sum_{j=1}^{J_{bn}} G_{ijm} \frac{\partial u_j}{\partial x_m} = \sum_{k=1}^{J_{bn}+J_{in}} \left\{ \alpha_k \left(\sum_{j=1}^{J_{bn}} H_{ij} \hat{u}_{jk} - \sum_{j=1}^{J_{bn}} G_{ij} \hat{q}_{jk} + c_i \hat{u}_{ik} \right) \right\} \tag{15}$$

where $G_{ijm} = G_{ij} n_m$. Substituting q from (14) into (13) yields:

$$\frac{\partial u_i}{\partial x_j} = -\sum_{m=1}^{J_{bn}} H_{im,j} u_m + \sum_p \sum_{m=1}^{J_{bn}} G_{imp,j} \frac{\partial u_m}{\partial x_p} + \sum_{k=1}^{J_{bn}+J_{in}} \left\{ \alpha_k \left(\sum_{m=1}^{J_{bn}} H_{im,j} \hat{u}_{mk} - \sum_{m=1}^{J_{bn}} G_{im,j} \hat{q}_{mk} + c_i \hat{u}_{ik} \right) \right\} \tag{16}$$

where $G_{imp,j} = G_{im,j} n_p$.

5 Solution procedures

There are two solution procedures applied depending on whether the considered node is on the boundary or in the interior of the domain.

5.1 Interior nodes

For interior nodes equations (15) and (16) are applied at each node. Since the potential and the partial derivatives are found from these two equations the system is closed and can be solved.

5.2 Nodes on the boundary

For nodes on the boundary the solution procedure applied by Natalini and Popov [2] has been applied.

When the source is in the interior and the integration is done over part of the boundary then instead of fluxes/normal derivatives, partial derivatives would be required in equations (15) and (16). They are obtained, as explained in [2], from the normal derivatives.

6 Numerical examples

The example used for verification of the approach is a 1-D Convection-Diffusion (C-D) problem with the following governing equation [3, 4]:

$$D\frac{d^2u}{dx^2}-\frac{d(V_x u)}{dx}=0 \tag{17}$$

A rectangular domain with length L and width W is considered. In order to produce equivalent 1-D results in a 2-D domain the following BCs were applied:

$$u(0,y)=U_0;\; u(L,y)=U_1;\quad \frac{\partial u}{\partial y}|_{y=W/2}=\frac{\partial u}{\partial y}|_{y=-W/2}=0 \tag{18}$$

The velocity field is defined as:

$$V_x=\ln\frac{U_1}{U_0}+k\left(x-\frac{1}{2}\right);\quad V_y=0. \tag{19}$$

The analytical solution of the above problem for $D=1\ m^2s^{-1}$ is given by:

$$u(x)=U_0\exp\left\{\frac{k}{2}x^2+\left(\ln\frac{U_1}{U_0}-\frac{k}{2}\right)x\right\} \tag{20}$$

The numerical results are obtained using the augmented thin plate spline (ATPS) [5] and $f = 1 + R$ [1] as DRM approximation functions. The results of the present approach are compared to the analytical results and the numerical ones obtained by the DRM-MD by Popov and Power [4].

In this example, three different meshes are generated in order to test the accuracy and the convergence of the method. The first mesh is generated with 95 nodes and 119 sub-domains. The second mesh is generated with 185 nodes and 229 sub-domains and is shown in Figure 3. The third mesh has 365 nodes and 449 sub-domains as shown in Figure 4. Two interior and two sub-domains on the boundary are marked with different colors. For each mesh two values of the parameter k (k = 10 and k = 40) and two approximation functions, the ATPS and $f = 1 + R$ are considered. The maximum velocity, equivalent to the Peclet number since D = L = 1, reach values 8.4 for k = 10 and 23.4 for k = 40.

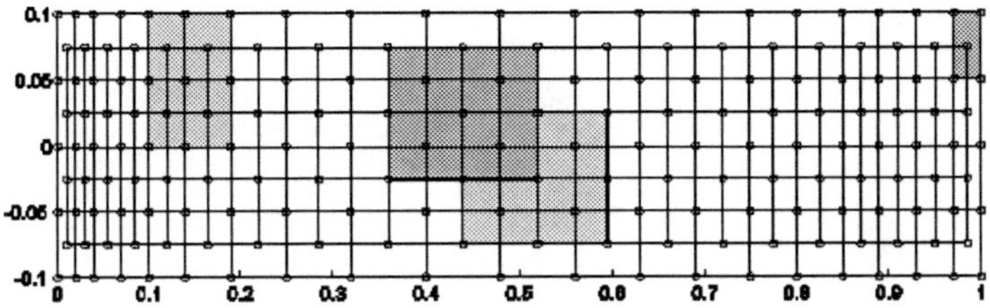

Figure 3: Mesh with 185 nodes used for the C-D problem.

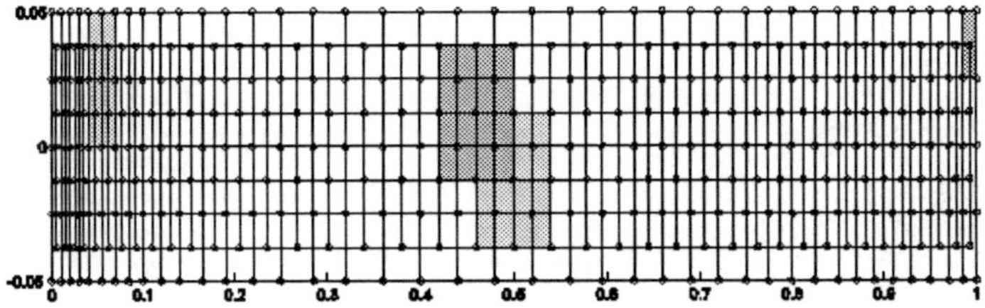

Figure 4: Mesh with 365 nodes used for the C-D problem.

The relative errors for the potential for the case $k = 40$ are plotted in Figure 5 for the ATPS and in Figure 6 for $f = 1 + R$. As can be seen in the figures the relative error decreases with the increase of number of nodes in the mesh, which confirms the convergence of the method. The coarse mesh with 95 nodes with ATPS had a maximum error of approximately 200% which is not shown in Figure 5.

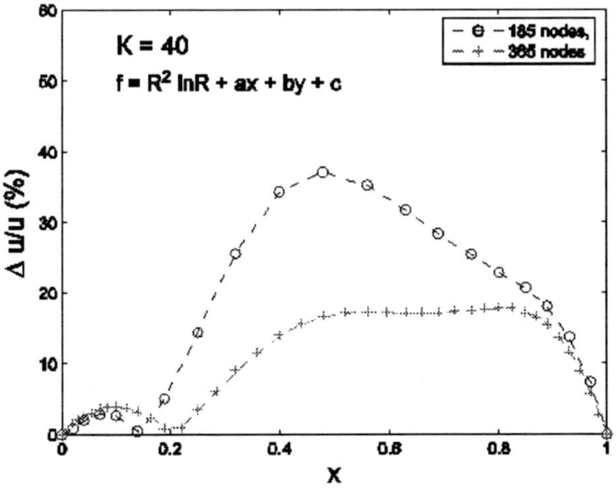

Figure 5: Relative error for the potential for the C-D problem (case $K = 40$, approximation function ATPS).

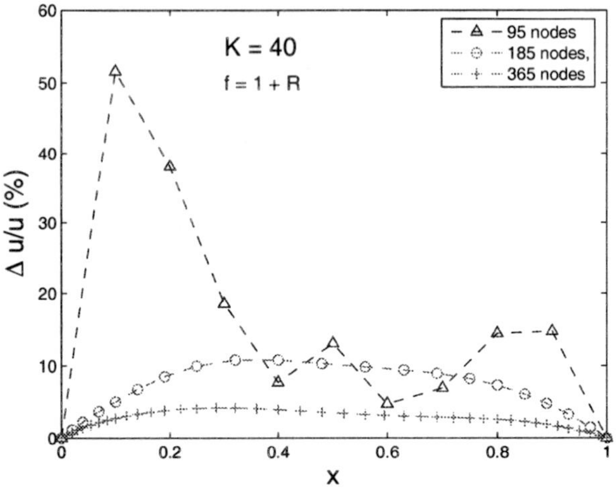

Figure 6: Relative error for the potential for the C-D problem (case $K = 40$, approximation function $f = 1 + R$).

Table 1: Relative error for the potential values at internal points obtained using the present method and the DRM-MD approach [4] for $k = 40$.

x	DRM-MD	$f = 1 + r$	ATPS
0.00	0.00000	0.00000	0.00000
0.02	1.60078	1.16151	0.93093
0.04	2.88296	2.26563	2.04848
0.07	5.16544	3.73569	2.93691
0.10	6.83983	5.05555	2.72869
0.14	9.38016	6.76309	0.48519
0.19	11.82482	8.51887	5.10446
0.25	13.46586	9.95951	14.41637
0.32	13.72012	10.76810	25.54530
0.40	12.41174	10.77337	34.33515
0.56	8.52606	10.46788	35.25292
0.75	3.98747	8.23023	25.47853
0.85	0.69131	6.09553	20.63368
0.93	0.52574	3.33171	13.70113
1.00	0.00000	0.00000	0.00000

In Table 1 the errors for the potential obtained by applying the DRM-MD and the proposed approach with the mesh with 185 and the two approximation functions are shown. It appears that the proposed approach with the approximation function $f = 1 + R$ produces better results than the other two formulations.

7 Conclusions

An overlapping BEM sub-domain technique has been developed with two different DRM approximation functions, ATPS and $f = 1 + R$. The proposed approach reduces some of the problems introduced by sub-domain techniques in the BEM, e.g., IBCs, which can introduce significant difficulties when implementing the numerical scheme, as well as singularities in the integrals arising from the formulation itself or distorted sub-domains due to complex geometries.

The formulation has been tested on a convection-diffusion problem with variable velocity for three different meshes with 95 nodes, 185 and 365 showing good accuracy and convergence with increasing refinement of the mesh. The approximation function $f = 1 + R$ has shown better accuracy than the ATPS.

The results obtained with the proposed approach have been compared to the DRM-D results obtained previously by Popov and Power [4], and showed higher accuracy than the DRM-MD when $f = 1 + R$ is used.

The proposed approach has been developed as an intermediate step towards a meshless integral equation approach which will be reported in the near future.

References

[1] Partridge, P.W., Brebbia C.A., Wrobel L.C. The dual reciprocity boundary element method, Southampton: Elsevier Applied Science; 1992.

[2] Natalini, B., Popov, V. (2004), An alternative approach for calculation of the first and higher order derivatives in the DRM-MD, *Engineering Analysis with Boundary Elements*, 28/1, 61–78.

[3] DeFigueiredo, D. B. Boundary Element analysis of conection-diffusion problems, Ph.D. Thesis, Wessex Institute of Technology, Southampton, U.K. 1990.

[4] Popov, V., and Power, H. The DRM-MD integral equation method: an efficient approach for the numerical solution of domain dominant problems, *Int. J. Numer. Meth. Eng.*, **44** (1999) 327–353.

[5] Golberg MA, Chen CS, The theory of radial basis functions applied to the BEM for inhomogeneous partial differential equations, Bound Elem Commun 1994; 5:57–61.

Section 6
Advanced computational techniques

Comparison of two solvers for the extended method of fundamental solutions

R. Gospavic[1], N. Haque[1], V. Popov[1] & C. S. Chen[2]
[1]*Wessex Institute of Technology, Ashurst Lodge, Southampton, UK*
[2]*Department of Mathematics, University of Southern Mississippi, Hattiesburg, MS 39406-5045, USA*

Abstract

In this work two solvers are compared for solution of the system of equations arising from the extended method of fundamental solutions. This method is a combination of the method of fundamental solutions (MFS) and the dual reciprocity method (DRM). The stationary and non-stationary elliptic partial differential equations with variable coefficients are considered. For the non-stationary case a finite difference time discretisation is used. Two sets of basis functions are used, a fundamental solution for the considered operator and a radial basis function (RBF) approximation for the particular solution, in particular the multiquadric (MQ) functions. A direct solver using a high number of digits and an indirect solver based on truncated singular value decomposition (TSVD) were used and compared. The method is tested for different values of the shape parameter in the MQ basis functions.

Keywords: meshless, method of fundamental solutions, elliptic partial differential equations, non-stationary problems.

1 Introduction

In recent times there has been growing interest in developing various meshless methods for the numerical solution of partial differential equations (PDEs) [1–3]. The major advantage of the meshless methods compared to the traditional numerical methods for solving PDEs is that they avoid the difficulties of meshing the domain and boundary, which is especially beneficial for 3D problems [2]. The use of meshless methods simplifies the geometrical representation of the problem domain and eliminates the need

WIT Transactions on Modelling and Simulation, Vol 47, © 2008 WIT Press
www.witpress.com, ISSN 1743-355X (on-line)
doi:10.2495/BE080191

for constructing mesh end element connectivity, instead the scattered nodes could be used without the need for connectivity [1]. The method of the fundamental solutions (MFS) uses the fundamental solution of the considered operator as the set of the basis functions for interpolation of the homogeneous solution [4–6]. To avoid the singularities the source points are placed outside of the problem domain on the fictitious boundary [7]. For the non-homogenous problem the method of the particular solutions (MPS) has been used to interpolate the non-homogeneous term by RBFs [8]. In contrast to the MPS where the homogenous particular solutions are obtained separately, in the present numerical method the particular and homogeneous solutions are combined and obtained simultaneously in one step [11, 12]. In this way the governing equations and boundary conditions are incorporated together in a single system of equations.

The two sets of basis functions are used simultaneously. The fundamental solution for the considered operator is used as the basis function for interpolation of the homogeneous solution and the MQ functions are used as RBF approximation for the particular solution [9]. For the non-stationary case the time discretization is used and the governing equation is solved by using time iterations. To obtain a high accuracy the method is tested for different values of the shape parameters. As the obtained system has a high condition number the two solvers are compared, the direct solver using a high number of digits and the TSVD solver.

2 The extended method of fundamental solutions (EMFS)

The following 2D time dependent boundary value problem (BVP) with appropriate boundary conditions (BC) will be considered:

$$\begin{aligned} &\Delta u(x,y,t)+\alpha(x,y)\cdot u(x,y,t)=\beta\cdot\frac{\partial u(x,y,t)}{\partial t}+f(x,y,t); \quad \vec{r}=(x,y)\in\Omega \\ &B\,u(x,y,t)=g(x,y,t); \quad \vec{r}=(x,y)\in\partial\Omega, \end{aligned} \tag{1}$$

where α(x,y), *f*(x,y,t) and *g*(x,y) are given functions, β is a constant and B is a first order operator, Ω is the 2D problem domain, and $\partial\Omega$ is the domain boundary. The non-homogeneous term could be interpreted as a source and the second term on the left side as a reactive term.

If time discretisation is performed and time derivative is approximated by finite difference the above equations could be rearranged in the following way:

$$\begin{aligned} &\Delta u(x,y,t_k)=F(x,y,u(x,y,t_k),u(x,y,t_{k-1})); \quad \Delta t=t_k-t_{k-1} \\ &F=-\alpha(x,y)\cdot u(x,y,t_k)+\beta\frac{u(x,y,t_k)-u(x,y,t_{k-1})}{\Delta t}+f(x,y,t_k), \end{aligned} \tag{2}$$

where Δt is the time step used.

The solution in the *k*-th time step is approximated as a sum of the particular and homogenous solutions in the following way [8, 10]:

$$u^k(\vec{r},t_k)=u_h^k(\vec{r},t_k)+u_p^k(\vec{r},t_k);\quad \Delta u_h^k=0;\quad \Delta u_p^k=F(\vec{r},u^k,u^{k-1});$$
$$u_h^k\approx\sum_{j=1}^{m}b_j^kG(|\vec{r}-\vec{\rho}_j|);\quad F(\vec{r},u^k,u^{k-1})\approx\sum_{i=1}^{n}a_i^k\varphi(|\vec{r}-\vec{r}_i|);\quad G(r)=4\pi\cdot\ln(r) \tag{3}$$
$$\Delta\hat{u}(|\vec{r}-\vec{r}_i|)=\varphi(|\vec{r}-\vec{r}_i|)\Rightarrow\quad u_p^k\approx\sum_{i=1}^{n}a_i\hat{u}(|\vec{r}-\vec{r}_i|),\quad \Delta u^k\approx F(\vec{r},u^k,u^{k-1}),$$

where $\varphi(r_i)$ are RBFs used for interpolation of the non homogeneous term, $G(\rho_j)$ is a fundamental solution for the 2D Laplace operator, u^{k-1} solution from previous time step, $\mathbf{r}_i$ are coordinates of the collocation points inside the domain Ω, $\boldsymbol{\rho}_j$ are coordinates of the source nodes on the fictitious boundary Γ, n is number of interior nodes inside Ω and m is the number of nodes on the boundary $\partial\Omega$ and in the same time the number of source points on the fictitious boundary Γ, as shown in Figure 1.

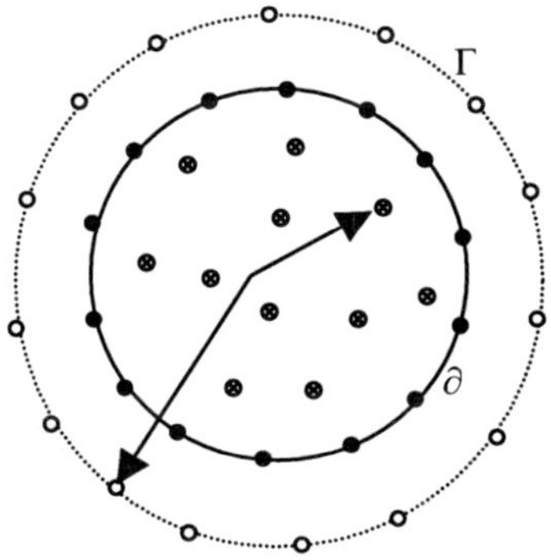

Figure 1: Arrangement of the internal, boundary and fictitious nodes.

By combining (2) and (3) the following expressions in k-th time step are obtained [8, 10]:

$$\Delta u^k=\sum_{i=1}^{n}a_i^k\Delta\hat{u}_i=\sum_{i=1}^{n}a_i^k\varphi_i=f(x,y,t^k)-\alpha(x,y)\cdot u^k+\beta\cdot\frac{u^k-u^{k-1}}{\Delta t}\Rightarrow$$
$$\sum_{i=1}^{n}a_i^k\Phi(|\vec{r}-\vec{r}_i|)+\sum_{j=1}^{n}b_j^k\Theta(|\vec{r}-\vec{\rho}_i|)\approx f(\vec{r},t^k)-\frac{\beta}{\Delta t}u^{k-1}(\vec{r}),\ \vec{r}=(x,y)\in\Omega \tag{4}$$

where Φ and Θ are given as:

$$\Phi(|\vec{r}-\vec{r}_i|)=\varphi(|\vec{r}-\vec{r}_i|)+\hat{u}(|\vec{r}-\vec{r}_i|)\cdot\left(\alpha(x,y)-\frac{\beta}{\Delta t}\right);$$
$$\Theta(|\vec{r}-\vec{\rho}_i|)=G(|\vec{r}-\vec{\rho}_i|)\cdot\left(\alpha(x,y)-\frac{\beta}{\Delta t}\right). \tag{5}$$

The boundary conditions in (1) are given by the following relation:

$$\sum_{i=1}^{n}a_i^k\cdot B\,\hat{u}(|\vec{r}-\vec{r}_i|)+\sum_{j=1}^{m}b_j^k\cdot B\,G(|\vec{r}-\vec{\rho}_j|)=g(x,y),\quad \vec{r}=(x,y)\in\partial\Omega \tag{6}$$

The collocation method is used and equation (4) is satisfied at n interior nodes and equation (6) is satisfied at m boundary nodes in each time step. In this way the following system of linear equations of order $(n + m) \times (n + m)$ is formulated for each time step:

$$\begin{bmatrix} \Phi(|\vec{r}_i - \vec{r}_j|) & \Theta(|\vec{r}_i - \vec{r}_j|) \\ B\hat{u}(|\vec{r}_i - \vec{r}_j|) & BG(|\vec{r}_i - \vec{r}_j|) \end{bmatrix} \cdot \begin{bmatrix} \tilde{a}^k \\ \tilde{b}^k \end{bmatrix} = \begin{bmatrix} \tilde{f}^k \\ \tilde{g} \end{bmatrix};$$
$$\tilde{a}^k = [a_1^k \cdots a_n^k]; \; \tilde{b}^k = [b_1^k \cdots b_m^k]; \tag{7}$$
$$\tilde{f}^k = \left[f(\vec{r}_1, t^k) - \frac{\beta}{\Delta t} u^{k-1}(\vec{r}_1), \cdots, f(\vec{r}_n, t^k) - \frac{\beta}{\Delta t} u^{k-1}(\vec{r}_n) \right]; \quad \tilde{g} = [g(\vec{r}_1), \cdots, g(\vec{r}_m)],$$

where $u^{k-1}(\mathbf{r}_i)$ is the solution from the k-1th time step in the interior node $\mathbf{r}_i$.

The MQ functions are used for interpolations of the non-homogeneous term and û function is expressed by the following equation [8]:

$$\hat{u}(r) = \frac{1}{9}(4c^2 + r^2) \cdot \sqrt{r^2 + c^2} - \frac{c^3}{3} \ln\left(c^2 + c \cdot \sqrt{r^2 + c^2}\right); \quad \varphi(r) = \sqrt{r^2 + c^2} \tag{8}$$

3 Numerical results

3.1 Example 1

In this example a stationary problem is considered, with non-homogeneous modified Helmholtz equation with a variable coefficient and appropriate boundary conditions of Dirichlet type, in the following form:

$$\begin{aligned} &\Delta u - x^2 y \cdot u = f(x, y); \quad (x, y) \in \Omega \\ &u(x, y) = \sin(y^2 + x) - \cos(y - x^2); \quad (x, y) \in \partial\Omega \\ &\Omega = \{(x, y) | x \in [0,1], \; y \in [0,1]\} \end{aligned} \tag{9}$$

If the non homogeneous term is given in the following form:

$$\begin{aligned} f(x, y) = &(1 + x^2 \cdot (4 + y)) \cdot \cos(x^2 - y) + 2\cos(x + y^2) + 2\sin(x^2 - y) \\ &- \sin(x + y^2) - (x^2 y - 4y^2) \cdot \sin(x + y^2) \end{aligned} \tag{10}$$

the exact solution is given by the second equation in (9). The rectangular domain with the nodes used is shown in Figure 2.

In Table 1 the numerical results for RMSE and max error for different shape parameters c obtained by using the TSVD and the direct solver as well as the condition number of the system matrix are presented

For this problem 49 interior nodes and 36 boundary and source nodes were used. The two different solvers used were: TSVD with cut off singular value equal to 10^{-13}, and a direct solver with 100 digits precision, in this case called arbitrary precision (AP), using the *Mathematica* software. The numerical error was estimated by using the root mean square error (RMSE) at testing nodes, which were uniformly arranged over the domain, and also by using the maximal relative error in a set of testing nodes. Twenty five testing nodes, arranged uniformly over the domain in a 5×5 grid were used. The RMSE is defined as follows:

$$RMSE = \sqrt{\frac{1}{Nt} \sum_{i=1}^{Nt} (\tilde{u}_i - u_i)^2}, \tag{11}$$

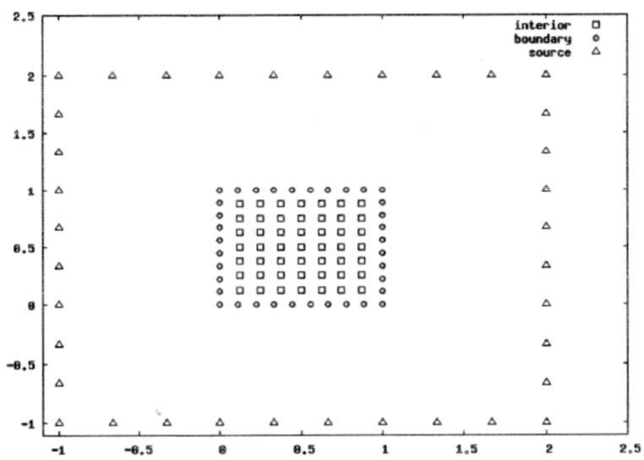

Figure 2: Problem domain with the arrangement of the interior, boundary and fictitious (source) nodes.

Table 1: Maximal relative error, RMSE and condition number of the matrix for the TSVD and the direct solver (AP).

c	% Max Error (TSVD)	% Max Error (AP)	RMSE (TSVD)	RMSE (AP)	Condition number
1	0.1418	0.1418	7.52E-006	2.11E-005	1.09E+028
2	0.1532	0.1502	2.75E-006	7.71E-006	2.96E+028
4	0.1110	0.0563	1.20E-005	2.58E-006	5.41E+029
6	22.676	0.0296	1.01E-003	1.91E-006	1.22E+033
8	482.839	0.0181	2.22E-002	1.90E-006	2.50E+036
10	1404.702	0.0133	5.50E-002	2.09E-006	9.21E+038
12	1274.686	0.0152	4.86E-002	2.47E-006	1.14E+041
14	7978.672	0.0260	0.27244	3.14E-006	6.73E+042
16	14148.345	0.0505	0.46112	4.32E-006	2.29E+044
18	26218.474	0.0961	1.207355	6.39E-006	5.16E+045
20	13318.156	0.1748	0.61568	9.97E-006	8.40E+046

where Nt is the number of testing points, $\tilde{u}_j$ and u_j are approximated and exact solutions in j^{th} node, respectively.

From the above results it can be concluded that the system is ill conditioned and that the condition number of the matrix is increasing with the increase of the shape parameter, however, the precision increases as well until a certain point. It was possible to show this behaviour of the solution only when using the AP solver, the TSVD solver failed to produce accurate results for higher values of the shape parameter, showing that it is more sensitive to the value of the shape parameter c.

3.2 Example 2

The following non-stationary problem with corresponding boundary conditions was considered:

$$\begin{aligned} \Delta u - x^2 y \cdot u &= \frac{\partial u}{\partial t} + f(x,y,t); \quad (x,y) \in \Omega \\ u &= \cos(\omega t) \cdot \left(\sin\left(y^2 + x\right) - \cos\left(y - x^2\right)\right); \quad (x,y) \in \partial\Omega \end{aligned} \tag{12}$$

The problem domain and arrangement of the nodes was the same as in the previous example. If the non homogeneous term is given in the following form:

$$\begin{aligned} & f(x,y,t) = \omega \sin(\omega t) \cdot \left(\left(\sin\left(y^2 + x\right) - \cos\left(y - x^2\right)\right)\right) + \\ & \cos(\omega t) \cdot \left(\left(1 + x^2 \cdot (4 + y)\right) \cdot \cos\left(x^2 - y\right) + 2\cos\left(x + y^2\right) \right. \\ & \left. + 2\sin\left(x^2 - y\right) - \sin\left(x + y^2\right) - \left(x^2 y - 4y^2\right) \cdot \sin\left(x + y^2\right)\right) \end{aligned} \tag{13}$$

the exact solution is given by the second equation in (12).

In Table 2 and 3 the numerical results for maximal relative error, RMSE and the condition number for different values of the shape parameter at different time instants are presented obtained by the direct solver from *Mathematica* software, representing each number with 100 digits and by TSVD solver respectively.

Table 2: Maximal relative error, RMSE and condition number of the matrix, for the various values of the shape parameter (c) at the different time instants for the non-stationary problem obtained by the AP solver.

C	1		5		10		15	
Con. No.	3.3E+14		1.48E+37		2.73E+37		5.72E+41	
Time [s]	Max rel. err %	RMSE	Max rel. err %	MSE	Max rel .err %	RMSE	Max rel err %	RMSE
0	0.1767	1.76E-5	0.0794	8.47E-6	0.0625	8.39E-6	0.0584	1.14E-5
0.5	0.2405	2.27E-5	0.1369	1.26E-5	0.1145	1.25E-5	0.1188	2.04E-5
1	0.2407	2.27E-5	0.1375	1.26E-5	0.1149	1.26E-5	0.1199	2.10E-5
1.5	0.2408	2.27E-5	0.1377	1.26E-5	0.1149	1.27E-5	0.1200	2.14E-5
2	0.2408	2.27E-5	0.1379	1.27E-5	0.1150	1.29E-5	0.1202	2.19E-5
2.5	0.2409	2.27E-5	0.1380	1.27E-5	0.1150	1.31E-5	0.1203	2.26E-5
3	0.2410	2.27E-5	0.1381	1.27E-5	0.1151	1.34E-5	0.1204	2.33E-5
3.5	0.2410	2.27E-5	0.1382	1.27E-5	0.1151	1.37E-5	0.1205	2.41E-5
4	0.2411	2.27E-5	0.1382	1.27E-5	0.1151	1.42E-5	0.1206	2.50E-5
4.5	0.2412	2.27E-5	0.1383	1.27E-5	0.1152	1.49E-5	0.1207	2.60E-5
4.9	0.2412	2.27E-5	0.1384	1.27E-5	0.1152	1.57E-5	0.1208	2.68E-5

The maximal relative error and RMSE are calculated at the same test nodes as in the previous example.

In Figures 3 and 4 the maximal relative error and RMSE as function of the time for different shape parameters obtained by AP solver are showed respectively.

From the above results for non-stationary case the same conclusion in respect to the shape parameter is valid, as for the stationary problem in Example 1. Increasing the shape parameter decreases the error of the numerical solution but the condition number increases making the system more ill conditioned. Again a direct solver with a high number of digits should be used to obtain an accurate

Table 3: Maximal relative error, RMSE and condition number of the matrix, for the various values of the shape parameter (c) at the different time instants for the non-stationary problem obtained by the TSVD solver.

c	1		5		10		15	
Con. No.	3.3E+14		1.48E+37		2.73E+37		5.72E+41	
Time	% error	RMSE	% error	RMSE	% error	RMSE	% error	RMSE
0.5	.23997	8.07E-6	1.4137	9.29E-5	8.259	1.40E-3	6485.801	0.2016
1	.24044	8.04E-6	6.5621	2.05E-4	165.043	1.70E-2	9558.103	0.3092
1.5	.24050	7.99E-6	2.6615	1.05E-4	90.0193	6.67E-3	13150.39	0.4376
2	.24057	7.92E-6	4.7890	1.87E-4	631.892	1.88E-2	5401.139	0.1298
2.5	.24064	7.84E-6	5.1626	1.95E-4	325.242	8.44E-3	33780.57	0.9882
3	.24070	7.73E-6	10.422	2.54E-4	350.234	7.81E-3	4042.020	0.2994
3.5	.24076	7.60E-6	6.3737	1.70E-4	1630.560	4.70E-2	56659.45	1.7605
4	.24083	7.45E-6	6.5225	1.72E-4	534.308	1.19E-2	47882.97	1.3022
4.5	.24091	7.29E-6	8.1751	2.20E-4	1440.991	3.90E-2	53565.09	1.6180
4.9	.24095	7.14E-6	8.2750	1.75E-4	795.872	2.73E-2	83151.63	2.3197

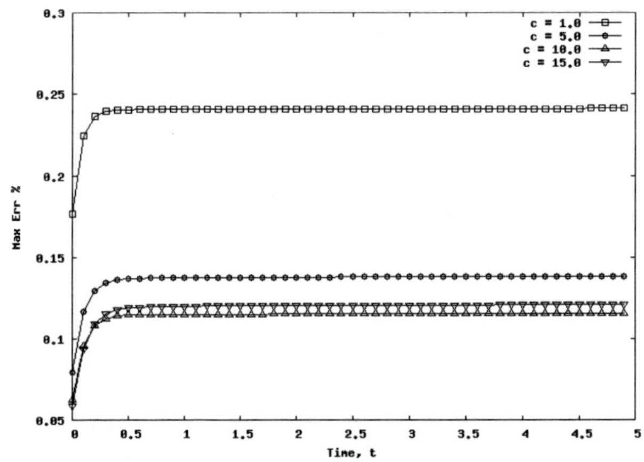

Figure 3: Maximal relative error versus time for different values of the shape parameter for results obtained using the AP direct solver.

solution that is less sensitive to the value of the shape parameter in the MQ function. In both cases the best results are obtained for shape parameter equal to 10.

4 Conclusions

A comparison is presented of results obtained using two different solvers with the meshless approach based on the method of particular solutions and the method of fundamental solutions. The stationary and non-stationary elliptic PDE is considered and a good agreement with the analytical solution is obtained, even

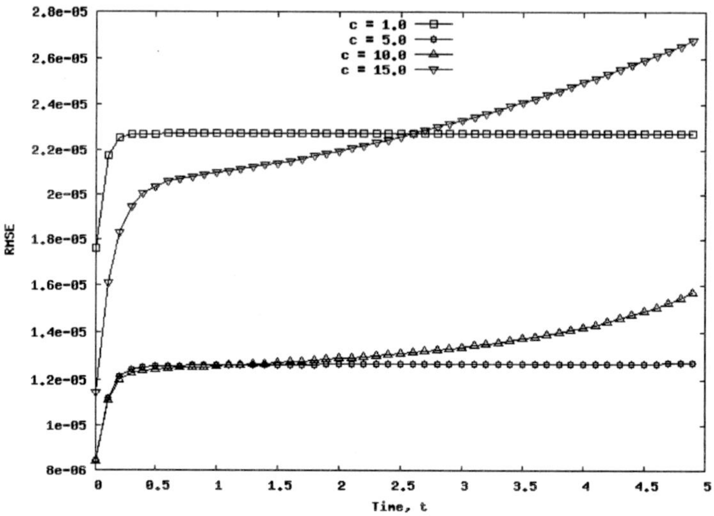

Figure 4: RMSE versus time for different values of the shape parameter for results obtained using the AP direct solver

for small numbers of interior and boundary nodes. In order to avoid singularities the source nodes for fundamental solutions are arranged outside the domain on a fictitious boundary. For approximation of the particular solutions MQ functions are applied. The accuracy of the method increases with the increase of the shape parameter, however, the condition number increases as well. The direct solver with a high number of digits, in this case 100 digits, was used in order to overcome the ill conditioning of the system matrix. An indirect solver based on the TSVD failed to solve the system for higher values of the shape parameter, showing higher sensitivity than the direct solver to the selection of the shape parameter. This results show that in some cases a large number of digits must be used in the solution of the system arising from this numerical approach.

References

[1] Jichun Li, Y.C. Hon, C.S. Chen, Numerical comparisons of two mesh less methods using radial basis functions, Engineering Analysis with Boundary Elements **26** (2002) 205–225.

[2] C.-S. Huang, C.-F. Lee, A.H.-D. Cheng, Error estimate, optimal shape factor, and high precision computation of multiquadric collocation method, Engineering Analysis with Boundary Elements **31** (2007) 614–623.

[3] M.A. Golberg, C.S. Chen, The method of fundamental solutions for potential, Helmholtz and diffusion problems. In: M.A. Golberg, editor. Boundary integral methods-numerical and mathematical aspects. Southampton: Computational Mechanics Publications; 1998. p. 102–76.

[4] Graeme Fairweather, Andreas Karageorghis, P.A. Martin, The method of fundamental solutions for scattering and radiation problems, Engineering Analysis with Boundary Elements **27** (2003) 759–769.

[5] G. Fairweather, A. Karageorgis, The method of fundamental solution for elliptic boundary value problems, Advances in Computational Mathematics, **9**, 69–95.

[6] C.S. Chen, Hokwon A. Cho, M.A. Golberg, Some comments on the ill-conditioning of the method of fundamental solutions, Engineering Analysis with Boundary Elements **30** (2006) 405–410.

[7] P.W. Patridge, B. Sensale, The method of fundamental solutions with dual reciprocity for diffusion and diffusion-convection using subdomains, Engineering Analysis with Boundary Elements **24** (2000) 633–641

[8] M.A. Golberg, C.S. Chen, Discrete Projection Methods for Integral Equations, Computational Mechanics Publications, Southampton.

[9] M.D. Buhmann, Radial basis functions: theory and implementations, University Press, Cambridge, 2003.

[10] P.W. Partridge, C.A. Brebbia, L.C. Wrobel The Dual Reciprocity Boundary Element Method. Computational Mechanics Publications, (1992), London

[11] C.S. Chen, C. M. Fan and J. Monroe, The method of fundamental solutions for solving elliptic partial differential equations with variable coefficients, submitted.

[12] H. Wang and Q.H. Qin A meshless method for generalized linear or nonlinear Poisson-type problems, Engineering Analysis with Boundary Elements, 30, 515–521, 2006.

Iterative 3D BEM solver on complex faults geometry using angular dislocation approach in heterogeneous, isotropic elastic whole or half-space

F. Maerten[1,2] & L. Maerten[1]
[1] *Igeoss, France*
[2] *University of Montpellier II, Dynamique de la Lithosphere, France*

Abstract

Based on the analytical solution of the induced displacement caused by a 3D angular dislocation, it is possible to construct closed polygonal loops with constant Burgers's vector, from which the stress is derived using linear elasticity in homogeneous, isotropic whole- or half-space. In this BEM code, each fault is discretized as a triangulated mesh, where mixed boundary conditions are prescribed.

Incorporate material heterogeneity is done by using triangulated interfaces made of dual-elements with prescribed continuity and equilibrium conditions. Each interface and fault can therefore have a complex 3D geometry with no gaps or overlaps between elements.

We use an iterative solver where the system of equations is decomposed at the element level, allowing a simple formulation of the boundary conditions for elements making a fault, and continuity/equilibrium conditions at dual-elements making an interface. It is shown that strict diagonal dominance can be achieved only if continuity and equilibrium conditions, for a given dual-element, are solved simultaneously. Using a Gauss-Seidel-like method, we consequently reduce the complexity while automatically taking care of the sparsity of the system. Moreover, using a Jacobi-like solver, we show that the resolution of the system can simply be parallelized on multi-core processors. Some comparisons with a 2D analytical solution and a 2D BEM code are presented.

Keywords: iterative solver, indirect method, heterogeneity, optimization, complex 3D geometry.

WIT Transactions on Modelling and Simulation, Vol 47,
www.witpress.com, ISSN 1743-355X (on-line)
doi:10.2495/BE080201

1 Introduction

In structural geology and geophysics, fault slip distributions play an important role for the induced stress perturbation. This slip distribution depends mainly on the geometrical configuration and boundary conditions along the faults, the remote boundary conditions as well as the constitutive behavior of the host rock. In particular, it was shown that the change in Young's modulus of the rock can affect the slip distribution in a non negligible way [1]. Therefore, it appears that heterogeneity cannot be ignored.

In 2D, Crouch and Starfield [2] proposed a method to add material heterogeneity into the Displacement Discontinuity Method (DDM). In 3D, it is possible to apply this method using the Okada's code [3] where fault surfaces are discretized into planar rectangular elements. However, this formulation necessarily introduces non-physical gaps and overlaps between adjacent elements which can perturb the solution [4].

In this paper, we use the analytical solution of an angular-dislocation in elastic, homogeneous, isotropic whole- or half-space, where the stress is derived using linear elasticity in homogeneous, isotropic whole- or half-space [5]. A boundary element method is then formulated by discretization of all complex 3D faults into triangular elements. This BEM code (called Poly3D [6]) is very similar to DDM [2], in which triangular elements of constant displacement discontinuity are employed. The advantage compare to Okada's code is that three-dimensional fault surfaces more closely approximate curviplanar surfaces and curved tiplines without introducing overlaps or gaps. Such formulation is very well suited to study faults interaction in 3D, since only faults surfaces have to be discretized (see for example [7–9] among others).

The addition of heterogeneous and isotropic materials is presented in this paper, and particularly an iterative method for solving the system is investigated.

2 BEM formulation

The BEM formulation employed here is derived from the analytical solution of an angular dislocation in 3D elastic whole- or half-space [5]. A triangular dislocation (or more generally a polygonal dislocation) with constant displacement discontinuity, or Burgers's vector b, can be constructed [6, 10] simply by superposition of six angular-dislocations (see Fig. 1). Mixed boundary conditions (BC) are prescribed, and when Neumann BC are specified, we have to solve for the unknown Burgers's components. After the system is solved, it is possible to compute anywhere, within the whole- or half-space, displacement, strain or stress at observation points as a postprocess.

Incorporate material heterogeneity (or general piecewise inhomogeneous bodies) requires discretizing an interface between two regions of different material properties into two triangulated meshes, one for each region. These two surfaces have the particularity to be perfectly coalescent, but with opposite normals (see Fig. 2). According to [2], continuity conditions on one side and equilibrium

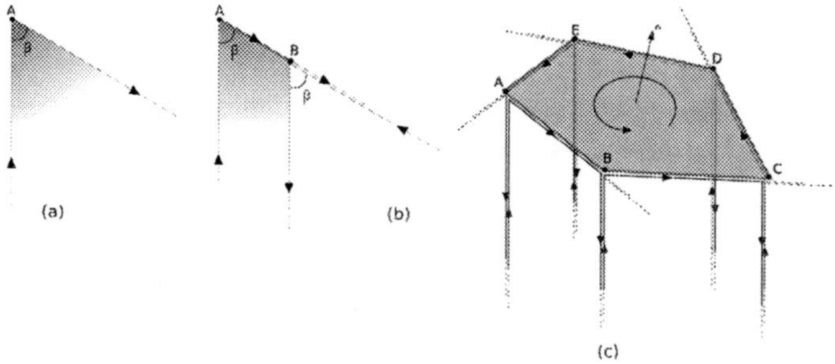

Figure 1: Construction of a polygonal element. (a) angular-dislocation, (b) a dislocation-segment made of two angular dislocations, and (c) a polygonal-element made of five dislocation segments of 10 angular-dislocations.

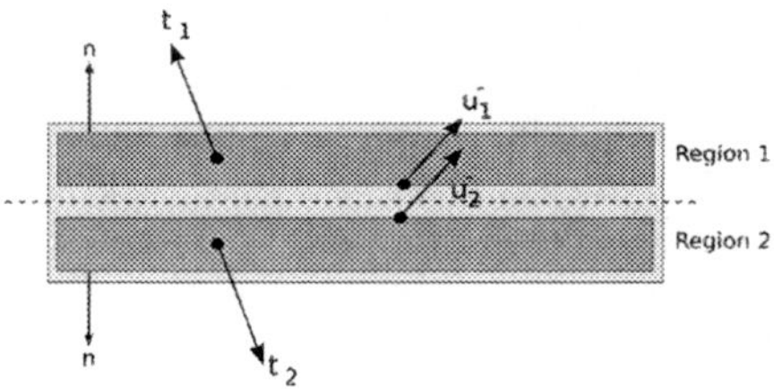

Figure 2: A dual-element part of an interface with equilibrium condition for tractions ($t_1 = -t_2$) and continuity conditions for displacement ($u_1^- = u_2^-$) on the negative side (in global coordinate system).

conditions on the other in element local coördinate system are applied [2] at interfaces (Fig. 2). Doing so requires particular attention for two main reasons. First, since we are in 3D, it is impossible to have two opposite local coordinate systems. Indeed, if x_i denotes the vectors of the first coordinate system ($i \in [0..2]$) and X_i the second, we have $X_0 = -x_0$, $X_1 = -x_1$ and $X_2 = X_0 \times X_1 = (-x_0) \times (-x_1) = x_0 \times x_1 = x_2 \neq -X_2$ (where $\times$ denotes the vector product). Consequently, we have to express continuity conditions in global coordinate system. The second reason is that since the displacement is discontinuous when going from one side of an element to the other, we have to make sure that U_{ee}^-, the self displacement influence matrix at the element center on the negative side, is correctly computed. Due to the machine precision, it is not always guaranteed, and consequently, we force the element center to be on the negative side by applying an infinitesimal translation of the center along the reversed normal.

Therefore, equilibrium and continuity conditions can be written in global coordinate system as:

$$\begin{cases} \overset{1}{\tau}_e = -\overset{2}{\tau}_e \\ \overset{1}{u}{}_e^- = \overset{2}{u}{}_e^- \end{cases} \tag{1}$$

where polygonal elements e in region 1 and 2 are perfectly coincident (they are called dual elements in the remaining of the paper).

A global system of equations is then built, which incorporates both equilibrium, continuity and boundary conditions:

$$\begin{bmatrix} \overset{1}{T}_I & \overset{1}{T}_F & \overset{2}{T}_I & \overset{2}{T}_F \\ -\overset{1}{D}{}_I^- & -\overset{1}{D}{}_F^- & \overset{2}{D}{}_I^- & \overset{1}{D}{}_F^- \\ \overset{1}{T}_I & \overset{1}{T}_F & 0 & 0 \\ 0 & 0 & \overset{2}{T}_I & \overset{2}{T}_F \end{bmatrix} \begin{Bmatrix} \overset{1}{b}_I \\ \overset{1}{b}_F \\ \overset{2}{b}_I \\ \overset{2}{b}_F \end{Bmatrix} = \begin{Bmatrix} 0 \\ 0 \\ \overset{1}{t}{}_F^0 \\ \overset{2}{t}{}_F^0 \end{Bmatrix} \tag{2}$$

where I and F stands for elements at interfaces and faults respectively. D^- and T represent the displacement influence matrix on the negative side and the traction influence matrix respectively, both of them in global coordinate system. The two first rows define the equilibrium and continuity conditions at interfaces, while the last two represent the classical boundary conditions applied to fault surfaces, and t_F^0 represents the initially prescribed traction boundary values for Neumann BC.

3 Iterative solver

In order to reduce the model complexity from $O(n^3)$ to $O(n^2)$ and to take advantage of the sparsity of the system (Eq. 2), an iterative solver is used.

The Burgers's vector solution at a regular boundary element e making a fault surface is given by:

$$b_e = T_{ee}^{-1}\left\{t_e^0 - \sum_{f \neq e} T_{ef} b_f\right\} \tag{3}$$

where T_{ef} denotes the traction influence matrix at the centroid of element e due to element f.

Computing the solution for a dual-element part of an interface is tricky since we are dealing with an iterative solver for which the convergence is guaranteed if and only if the system is strict diagonal dominant [11], i.e. $\forall i, \forall j \neq i, |a_{ii}| > |a_{ij}|$. As an element e_1 and its dual part e_2 have the same geometry (only the orientation changes), $T_{e_1e_1}$ and $T_{e_2e_2}$ have the same diagonal values, and strict diagonal dominance is not honored (the same apply for D^- matrices). Therefore, equilibrium condition in region R_1 and continuity condition in region R_2 have to be solved simultaneously for a given dual-element. This leads to the following coupled elemental system:

$$\begin{Bmatrix} \overset{1}{b}_e \\ \overset{2}{b}_e \end{Bmatrix} = \begin{bmatrix} \overset{1}{T}_{ee} & \overset{2}{T}_{ee} \\ -\overset{1}{D}{}_{ee}^- & \overset{2}{D}{}_{ee}^- \end{bmatrix}^{-1} \begin{Bmatrix} -\sum_{f \neq e} \overset{1}{T}_{ef}\, b_f - \sum_{g \neq e} \overset{2}{T}_{eg}\, b_g \\ \sum_{f \neq e} \overset{1}{D}{}_{ef}^-\, b_f - \sum_{g \neq e} \overset{2}{D}{}_{eg}^-\, b_g \end{Bmatrix} \tag{4}$$

Eqs. (3) and (4) are then solved using a Gauss-Seidel procedure as follow:

while $\epsilon >$ tolerance do
for all element e in region R_1
if e is not part of an interface
let $t =$ initial traction vector of e
for all element f in region R_1 except e
$t = t - T_{ef} b_f$
end for
Use Eq. (3)
update ϵ and set b to e
else
let $t = 0$ and $u^- = 0$
for all element f in region R_1 except e
$t = t - T_{ef} b_f$ and $u^- = u^- + D^-_{ef} b_f$
end for
for all element g in region $\bar{R}_2$ except e
$t = t - T_{eg} b_g$ and $u^- = u^- - D^-_{eg} b_g$
end for
Use Eq. (4)
update ϵ and set $\overset{1}{b}$ to e and $\overset{2}{b}$ to dual(e)
end if
end for
end while

This algorithm automatically takes care of the sparsity of the system, while allowing large model computation since practically no memory allocation is required.

4 Results

In order to check the validity of the formulation, we first compare the 2D analytical solution [2] of an annulus (with $\nu_1 = 0.25, G_1 = 1$) inside a circular hole in a large plate ($\nu_2 = 0.25, G_2 = 0.5$) (Fig. 3(b)), with a corresponding 3D model (Fig. 3(a)). The hole is subjected to an internal pressure of 0.001 Pa, and the computed normalized σ_3 and σ_1 along the bold lines are compared. In order to avoid boundary effects, the tubes, defining the annulus and the hole in 2D, are extended far from the zone of interest, materialized by the squared observation plane (Fig. 3(a)). It can be seen that the computed values along the bold line in Fig. 3(a) match the analytical solution.

Comparison with an existing 2D BEM code (Fig. 4) is described in [1]. The model is composed of a penetrating fault inside an inclusion, and various configurations are computed where the ratio of the Young's modulus in the inclusion to that of the host material, was set to 1, 0.1 and 10. It is subjected to a uniform remote unit shear stress σ_{yx} of 1 MPa. The corresponding 3D model is composed of two orthogonal vertical planar surfaces, one for the interface defining the inclusion and

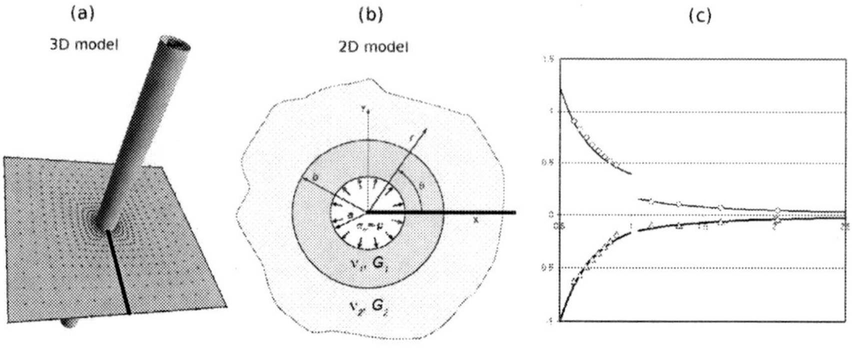

Figure 3: Comparison with a 2D analytical solution. (a) 3D model configuration, (b) 2D model, (c) comparison of normalized σ_3 (triangles) and σ_1 (circles), where solid lines are for the numerical solutions.

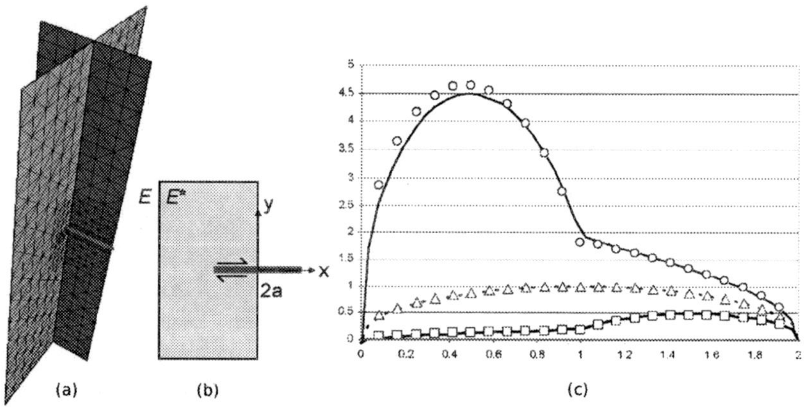

Figure 4: Comparison with a 2D BEM code. (a) 3D model configuration, (b) 2D model, (c) comparison of the computed dip-slip component onto the fault (along the bold line in (a) and (b)). Triangles is for $E^*/E = 1$, circle for $E^*/E = 0.1$, and squares for $E^*/E = 10$.

the other for the penetrating fault. The other five sides of the box inclusion are not taken into account. The bold lines in Fig. 4(a) and 4(b) represents the x-axis of Fig. 4(c). Computed dip-slip component are reported onto the Y-axis. Again, the results match the 2D BEM solutions.

5 Optimizations

Even if this iterative algorithm is in $O(n^2)$, it is slower than the direct solver, since at each iteration and for each element, the influence matrices (traction and displacement) due to all other elements have to be re-calculated.

Table 1: Speed comparison (in seconds) for different model sizes, performed on Intel Xeon 2Ghz with 8 cores running Linux Ubuntu. "LU" is the classical direct solver of the whole system, "Iter" is the iterative solver with no memory allocation, "Alloc" is the iterative solver with memory allocation, and "Thread" is the parallelized version on 2 or 8 threads. Models marked "x" where stopped before the end.

Model size	LU	Iter	Alloc	Thread 2	Thread 8
2200	49	380	9	8	6
4200	466	1018	46	26	8
9000	1145	5600	205	116	32
16000	x	x	1200	690	205

5.1 Bufferized elemental matrices

We can highly increase the speed if we store elemental matrices T_{ef}, D^-_{ef}, T_{eg} and D^-_{eg} from Eqs. (3) and (4) for each element e. Of course, if we reach the maximum available RAM (Random Access Memory), the remaining matrices have to be computed "on the fly" in order to avoid memory "swapping".

Furthermore, since the global system is never constructed nor inverted, there is no cumulative roundoff error when using elemental matrices with floating precision (4 bytes) instead double (8 bytes), allowing larger models to be computed.

Table 1 gives some examples of the gain of speed using a direct LU solver and the iterative solver with and without allocation.

5.2 Parallelization on multi-core processors

The iterative solver allows taking advantage of the new multi-core processors architecture, by parallelizing the computation onto different threads using the cross-platform package [12] (where the number of threads k is defined by number of processors on the mother board times the number of cores for each processor). At the beginning, the system is split into k sub-systems (the decomposition technique is irrelevant), one for each thread. Consequently, k iterative solvers are run in parallel. There is no need to update the communication between each sub-process at the end of each iteration since they share the same model memory. The only constraint is to update the new displacement onto the elements at the end of each iteration by waiting for each thread to finish its job in order to avoid read/write conflicts. This is simply achieved by using a Jacobi procedure instead of the Gauss-Seidel one. Table 1 gives an overview of the computation time for $k = 2$ and $k = 8$.

6 Conclusions

Indirect BEM techniques appear to be an advantageous way of modeling stress perturbation around faulted area in whole- or half-space since only the fault surfaces have to be discretized as boundary elements. Implementing material heterogeneity requires little effort and the use of an iterative solver have a great impact in terms of speed for the system resolution and memory consumption. The Jacobi version of the iterative solver also permits parallelization on multi-core processors in a simple and efficient way.

Since each dual element generates six unknowns into the system, it is necessary to reduce the complexity, which is part of our current research.

References

[1] Burgmann, R. & Pollard, D.D., Inverting for slip on three-dimensional fault surfaces using angular dislocations. *Journal of Structural Geology*, **16(12)**, pp. 1675–1690, 1994.

[2] Crouch, S.L. & Starfield, A.M., (eds.) *Boundary element methods in solid mechanics*. Unwin Hyman: London, 1983.

[3] Okada, Y., Surface deformation due to shear and tensile faults in a half-space. *Bulletin of the Seismological Society of America*, **75**, pp. 1135–1154, 1985.

[4] Maerten, F., Resor, P.G., Pollard, D.D. & Maerten, L., Inverting for slip on three-dimensional fault surfaces using angular dislocations. *Bulletin of the Seismological Society of America*, **95**, pp. 1654–1665, 2005.

[5] Comninou, M. & J., D., The angular dislocation in a half space. *Journal of Elasticity*, **5(3)**, pp. 203–216, 1975.

[6] Thomas, A.L., *Poly3d: a three-dimensional, polygonal element, displacement discontinuity boundary element computer program with applications to fractures, faults, and cavities in the earth's crust*. Master's thesis, Stanford University, 1993.

[7] Maerten, L., Willemse, E.J.M., Pollard, D.D. & Rawnsley, K., Slip distributions on intersecting normal faults. *Journal of Structural Geology*, **21**, pp. 259–271, 1999.

[8] Maerten, L., Variation in slip on intersectiong normal faults: Implications for paleostress inversion. *Journal of Geophysical Research*, **105(25)**, pp. 553–565, 2000.

[9] Muller, J.R., Aydin, A. & Maerten, F., Investigating the transition between the 1967 mudurnu valley and 1999 izmit earthquakes along the north anatolian fault with static stress changes. *Geophysics Journal International*, **154**, pp. 471–482, 2003.

[10] Jeyakamuran, M., Rudnicki, J.W. & Keer, L.M., Modeling slip zones with triangular dislocation elements. *Bulletin of the Seismological Society of America*, **82**, pp. 2153–2169, 1992.

[11] Golub, G.H. & Loan, C.F.V., (eds.) *Matrix computation*. ohns Hopkins Studies in the Mathematical Sciences, 1996.

[12] Trolltech, Qtconcurrent, http://www.trolltech.com.

Adaptive cross approximation based solver for boundary element method with single domain in 3D

T. Grytsenko & A. Peratta
Wessex Institute of Technology, Southampton, UK

Abstract

The approach presented in this paper is based on the Adaptive Cross Approximation (ACA) applied to BEM-derived matrices. The ACA method is purely algebraic and does not require information on the kernel. In this way, it offers an interesting alternative to the well established Fast Multipole Method. The algorithm uses a hierarchical approach for the matrix storage and manipulation. This paper describes the assembly approach for the usual H and G matrices coming from BEM equations into a hierarchical coefficient matrix A_H. The efficiency of the developed scheme is tested for a potential problem involving a 3D unitary cube, and then the method is applied to an electrostatic problem involving an anatomic model of the human body. This 3D example consisting of 80 000 degrees of freedom could easily be solved in a low specification PC.
Keywords: boundary elements, adaptive cross approximation, hierarchical matrices, linear systems of equations.

1 Introduction

This paper considers some technical aspects related to implementation of the Adaptive Cross Approximation (ACA) algorithm [1] to general potential problems solved by the collocation Boundary Element Method (BEM). The great advantage of the ACA algorithm is that it is purely algebraic and does not need to operate with the kernel. The algorithm uses a hierarchical matrix storage approach [2] where the matrix is split into many blocks classified into two categories, weakly and strongly coupled. The former are off-diagonal blocks which represent remote interactions between source points and field elements, and therefore can be approximated by low-rank matrices. For this purpose the ACA algorithm is used [2]. These blocks

WIT Transactions on Modelling and Simulation, Vol 47, © 2008 WIT Press
www.witpress.com, ISSN 1743-355X (on-line)
doi:10.2495/BE080211

are stored in a special Rk-format [2]. The latter blocks describing close interactions between source points and field elements are stored without any changes in a full-matrix format [2].

The novelty of this paper consists in a formulation of the approach allowing to assemble both H and G matrices from BEM formulation (see Section 2) within only one hierarchical matrix A_H. Two different ways to implement the approach for existing BEM formulations are considered. This paper demonstrates the efficiency of the approach with more than 80 000 degrees of freedom (DOF) and reports its numerical properties. This paper is organised as follows: in Section 2, the BEM method formulation for a general electrostatic problem is presented; in Section 3, the assembly scheme applied to a simple example is described; Section 4 suggests different ways to implement the approach within existing BEM formulations; Section 5 presents the numerical results of the application to electrostatic problem; Section 6 elaborates the concluding remarks.

2 Boundary element method formulation

Consider solving the Laplace equation for the unknown scalar field $u(\mathbf{x})$ given by:

$$\nabla^2 u(\mathbf{x}) = 0, \quad \mathbf{x} \in \Omega \subseteq R^3 \tag{1}$$

where Ω is the integration domain with boundary $\Gamma = \partial(\Omega)$ of outward unit normal $\hat{\mathbf{n}}$, and proper boundary conditions are applied to Γ, i.e. Dirichlet or Neumann type. Then, the boundary integral formulation for eq. (1) can be expressed in the following way [3]:

$$c_i u(\mathbf{x}_i) + \int_\Gamma q^*(\mathbf{x}, \mathbf{x}_i) u(\mathbf{x}) d\Gamma - \int_\Gamma u^*(\mathbf{x}, \mathbf{x}_i) q(\mathbf{x}) d\Gamma = 0, \tag{2}$$

where q is the normal derivative of u in $\hat{\mathbf{n}}$ direction, u^* is the Green's function of Laplace equation such that $\nabla^2 u^* = 0$, q^* its normal derivative in $\hat{\mathbf{n}}$ direction, and c_i is the self-interaction coefficient. In 3D problems u^* and q^* become: $u^* = 1/(4\pi r)$ and $q^* = -\mathbf{r} \cdot \hat{\mathbf{n}}/(4\pi r^3)$, respectively, where $\mathbf{r} = \mathbf{x} - \mathbf{x}_i$ and $r = |\mathbf{r}|$ is the distance between the field ($\mathbf{x}$) and source ($\mathbf{x}_i$) points.

In order to solve eq. (2), Γ is discretised into N_e constant triangular boundary elements Γ_e. Thus, the discretised boundary integral equation becomes:

$$c_i u_i + \sum_{j=1}^{N_e} h^{(j)} u_j - \sum_{j=1}^{N_e} g^{(j)} q_j = 0, \tag{3}$$

where u_j is the average potential in the j-th element, q_j is the mean normal flux in the centroid of j-th element, and $q^{(j)}$ and $h^{(j)}$ are the following BEM integrals:

$$g^{(j)} = \frac{1}{4\pi} \int_{\Gamma_j} \frac{1}{r} d\Gamma_j \tag{4}$$

$$h^{(j)} = \frac{1}{4\pi} \int_{\Gamma_j} \frac{\mathbf{r} \cdot \mathbf{n}}{r^3} d\Gamma_j \tag{5}$$

The assembly scheme consists in appending one equation (2) per each selected source point $\mathbf{x}_i$, to the global system of equations ($\mathbf{A}\ \mathbf{x} = \mathbf{b}$), where $\mathbf{A} \in R^{n \times m}$ contains the coefficients $h^{(j)}$ and $g^{(j)}$, $\mathbf{x}$ is a 1-column array with the unknown u and q, $\mathbf{b}$ is the right hand side 1-column array formed by the boundary conditions. The matrix A will be presented in hierarchical format using the techniques described in [2]. Each admissible block [2] is approximated using the ACA algorithm [1].

3 The assembly scheme

In order to take advantage of the efficiency provided by H-matrices in conjunction with ACA approximation it is not only necessary to build the structure of the matrix A_H from the equation $Ax = b$ but also to compute all the entries and assemble them in an appropriate way. There are two radically different ways to perform this operation: for the case when all the entries of matrix A have to be computed in advance and for the case when only several rows and columns of the initial matrix have to be calculated. The former approach has been referred to as fully-pivoted ACA algorithm in [1], while the latter as the partially-pivoted version of ACA algorithm [1].

Both approaches are based on a hierarchy of blocks in the coefficient matrix A_H called supermatrix [2]. Each block represents one of three cases: Full-block, Rk-block or a supermatrix of a lower level. The Full-blocks can not be approximated and therefore are computed and stored without any changes, while the Rk-blocks are approximated an stored in a factorized form according to the following expression:

$$B_{RK} = \sum_{i=1}^{k} UV^T \tag{6}$$

where k is the rank of the approximated block, $U \in R^{nxk}$ and $V \in R^{mxk}$, n and m are number of rows and columns in an Rk-block.

If the block under analysis is a supermatrix, it has to be further divided into either Full- or Rk-block. Due to the structure of A_H, the assembly approach is a recursive process.

The only difference with the partially pivoted ACA approach [1] is that the Rk-blocks can be generated on the fly during the assembly according to the algorithm of partially pivoted ACA [1]. The assembly scheme for the Full-blocks will remain the same: the matrix entry A_{ij} corresponds to the contribution of boundary element j to i-th BEM equation. Therefore A_{ij} element involves the source point x_i and the DOF associated with j BE.

It is worth mentioning that the coefficients of H and G matrices can be computed using different numerical algorithms. The computation of H and G is performed by an external function that gives as an output the values of corresponding h and g coefficients. Hence, the whole algorithm for the computation of sub-block entries can be represented in the following way:

$$Block[i][j] = \begin{cases} h_{ij} & \forall i \neq j \text{ and } bc[j] = 1 \\ g_{ij} & \forall bc[j] = 0 \\ 1/2 & \forall i = j \text{ and } bc[j] = 1 \end{cases} \qquad (7)$$

where $i = 1 : rows$, $j = 1 : cols$, vector bc contains the information about boundary condition (BC), i.e. it contains 0 or 1 for the corresponding degree of freedom if appropriately the potential or normal flux are known:

$$bc[j] = \begin{cases} 0 & \text{if DOF } j \text{ has Dirichlet-type BC} \\ 1 & \text{if DOF } j \text{ has Neumann-type BC} \end{cases} \qquad (8)$$

The i-th row of the right hand side vector RHS can be computed as follows:

$$RHS[i] = \sum_{j=1/bc[j]=1}^{cols} g_{ij}q[j] - \sum_{j=1//bc[j]=0}^{cols} h_{ij}u[j] \qquad (9)$$

where $i = 1 : rows$, vector u contains appropriate values for known potentials while vector q stores known normal fluxes. If $bc[i] = 0$, then the $RHS[i]$ has to be updated as follows:

$$RHS[i] = RHS[i] - u[i]/2 \qquad (10)$$

3.1 An example

This section illustrates the assembly scheme with a worked example. The problem represents a Laplacian equation in a 3D unitary cube (ie. of dimensions $1 \times 1 \times 1$) with conductivity $k = 1$ (see Figure 1). The model has mixed boundary conditions: Dirichlet for the potentials and Neumann for the normal fluxes. The boundary has been meshed with constant triangular elements yielding 12 discontinuous freedom nodes which are located in the centroid of each element (see Figure 2). Taking into account the boundary conditions, the model yields $N = 12$ unknowns: 4 normal fluxes and 8 potentials. A cluster tree [2] for this model is obtained with the help of bisection clustering technique [4] using the maximal number of indices per node $n_{\min} = 4$ [2]. Its clusters and the corresponding H-matrix A_H are shown in Figure 2. The number in each block represents its rank, shaded blocks represent the Full-blocks while the white ones are Rk-blocks. A_H a square matrix $A_H = [12 \times 12]$ which has 16 blocks where only 2 of them satisfy the admissibility condition [2]. Those two blocks are admissible and stored in Rk-format [2]. For the sake of simplicity, the assembly scheme of only two blocks, i.e. $B_F = [(1, 2) \times (1, 2)]$ and $B_{RK} = [(1, 2) \times (11, 12)]$ is considered. The block B_F is represented by

$vertex/el_{id}$	el_1	el_2	el_{11}	el_{12}
x_1	(0,1,0)	(0,0,0)	(1,1,0)	(1,1,0)
x_2	(0,0,0)	(1,0,0)	(1,1,1)	(0,1,0)
x_2	(0,0,1)	(0,0,1)	(1,0,1)	(1,1,1)

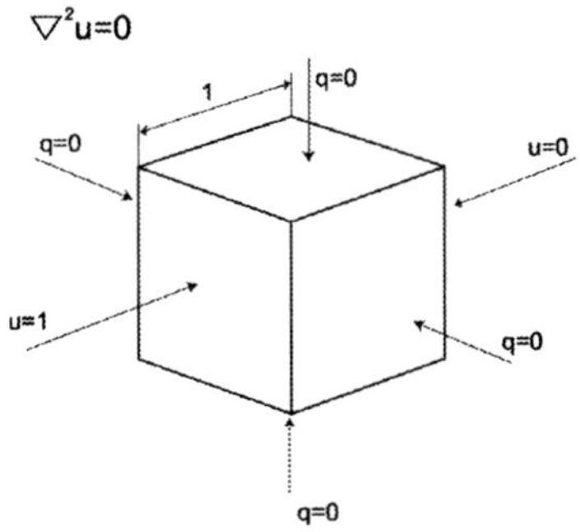

Figure 1: Model “cube”.

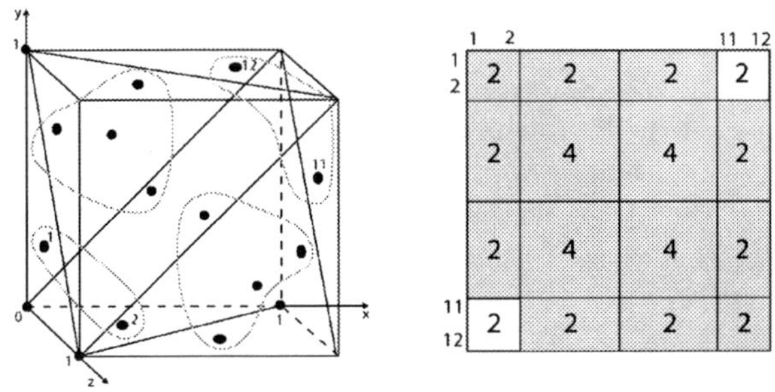

Figure 2: A collection of clusters for a test problem and the corresponding H-matrix.

Full-matrix format while the block B_{RK} is an Rk-block. There are several vectors storing the auxiliary information about the model. For example the vector

$$x_s = [(0, 0.33, 0.33); (0.33, 0, 0.33)], \tag{11}$$

consists of the 3D coordinates of the source points corresponding to the first and second row in the super-matrix A_H, i.e. row number 1 corresponds to the source point with 3D coordinates $(x, y, z) = (0, 0.33, 0.33)$ whereas the row number 2 corresponds to $(0.33, 0, 0.33)$; three vectors x_1, x_2 and x_3 that consist of the 3D coordinates of the vertices of triangular field elements corresponding to each column in the super-matrix A_H (see Table 3.1). In Table 3.1 the elements el_1, el_2, el_{11} and el_{12} correspond to the elements indicated in Figure 2 (the id of the field element coincide with the id of the DOF). The bc vector storing boundary conditions looks as follows:

$$bc = [0, 1, 0, 1], \tag{12}$$

i.e. the potential is known for the first element corresponding to column number 1. The vector of known potentials looks as follows:

$$u = [1, und, 0, und], \tag{13}$$

i.e. known potential for the first element is equal to 1, however, the potential for the element number 2 is not known and therefore represented by *und* which means undefined value. The vector of known normal fluxes stores the following data:

$$q = [und, 0, und, 0], \tag{14}$$

i.e. known normal fluxes for second and twelfth elements are both equal to 0. Having vectors (11–14) in mind and corresponding to eq. (7), the entries of block B_F and B_{RK} have to be computed as follows:

$$B_F = \begin{bmatrix} 0.1915 & -0.1148 \\ 0.0768 & 0.5 \end{bmatrix}, \quad B_{RK} = \begin{bmatrix} 0.0346 & -0.0324 \\ 0.0385 & -0.0304 \end{bmatrix} \tag{15}$$

Once the B_{RK} block is computed, it can be approximated within two iterations following the algorithm of fully pivoted ACA presented in [1]. As it has been mentioned in eq. (6), the Rk-blocks are presented in factorised form using the vectors V and U. At the end of first iteration, they look as follows:

$$V = [0.0346\ 0.0385], \quad U = [1\ -0.7901] \tag{16}$$

The next iteration leads to the final approximation as it reaches the maximal rank of block B_{RK}:

$$V = [0.0346\ 0.0385\ -0.00503\ 0], \quad U = [1\ -0.7901\ 0\ 1] \tag{17}$$

As a result of second iteration, the matrix approximated without any error, i.e. the initial block and the block after approximation are absolutely the same. It is possible to check this fact using the expression (6). In this particular example, the rank of the approximated block is equal to rank of the initial matrix. However, when the rank is fairly high the difference may reach hundreds of times economizing the memory consumption and reducing the complexity of matrix-vector multiplications.

4 Different ways to implement an ACA-based solver for existing BEM formulations

There are two ways to implement ACA-based solver for the solution of LSE coming from BEM method:

(i) As a "grey-box" solver, in which the solver takes as an input: the coefficient matrix A, the right hand side of the equation $Ax = b$ and the information about geometry, which could be provided by an existing BEM code;

(ii) or integrated within the BEM strategy where the assembly scheme is embedded within the BEM code. In this case A matrix has to be assembled according to its hierarchical representation during the evaluation of H and G matrix coefficients in the BEM formulation.

The first approach needs to store two copies of the matrix A at the same time, i.e. one original coefficient matrix A and its hierarchical representation A_H, i.e. the number of entries N_{ST} can be estimated as follows:

$$N_{ST} = NM + \sum_{i=1}^{N_{RK}} k_i(n_i + m_i) + \sum_{i=1}^{N_F} n_i m_i \tag{18}$$

where $N = rows$ and $M = cols$ are the number of rows and columns in the original matrix A, N_{RK} and N_F are the number of Rk and Full blocks appropriately, n_i and m_i are appropriately the number of rows and columns in block i of the coefficient H-matrix A_H. The term NM from the eq. (18) is reduced with the help of hierarchical matrix in conjunction with ACA approach. As a practical rule of thumbs for current PC systems (in this case Pentium iii, 1 GHz, 2 Gb RAM), if the model under analysis is fairly complex, the only way to solve it within reasonable memory space is to read the corresponding entries from the initial matrix A block by block. In this case, the memory capacity can be reduced up to:

$$N_{ST} = \sum_{i=1}^{N_{RK}} k_i(n_i + m_i) + \sum_{i=1}^{N_F} n_i m_i \tag{19}$$

However, the CPU time T_{CPU} increases according to:

$$T_{CPU} = N_B(T_S + T_R + T_P) \tag{20}$$

where N_B is the number of blocks to be read, T_S is the time required to find the data in a file, T_R is a time needed to access the data and T_P is a CPU time to process the entry. If the number of blocks N_B is too high, the total CPU time T_{CPU} increases dramatically due to the fact that the time to find a corresponding entry in the original matrix increases non-linearly even for advanced searching techniques. On the other hand, if N_B is small, the CPU time is reasonable but the algorithm risks to go beyond the memory limits as the storage requirement for each block needed to be read is $O(N^2)$. As an example, if the whole previously assembled matrix has to be read, the model consisting of nearly 12 000 degrees of freedom needs about 1.3 Gb of RAM according to (18) and about 1 hour to be assembled. However, if it is read block by block, the total memory capacity needed for storing the A_H matrix is just about 200 Mb and takes about 3 hours to be assembled.

Another way to overcome the problem with the memory requirement mentioned above is to use the ACA within the computation of the coefficients of H-matrix A_H. In this case, the method does not need to store any additional information and can compute the coefficients on the fly. The storage requirement for this approach is as shown in the eq. (19). The total CPU time can be evaluated as follows:

$$T_{CPU} = N_B(T_C + T_P) \tag{21}$$

where T_C is a CPU time to compute the entries which is much less than the time to find the needed entry in a file when the large scale models are involved. For example, the model with nearly 12 000 degrees of freedom took about 3 hours to be assembled (on a Pentium iii, 1 Ghz, 2 Gb RAM) reading the coefficients block by block, while for the case when the entries are computed on the fly, it takes just about 20 minutes on the same PC. The memory requirement is same as for the reading the matrix block by block and for this particular example is about 200 Mb.

The partially-pivoted ACA allows to generate the Rk-blocks on the fly without storing the intermediate block [1]. Moreover, it does not need to compute all the entries. Although the partially pivoted ACA is slightly more difficult to implement in comparison with the fully pivoted version, it works faster as it does not need to compute all entries of the H-matrix A_H beforehand. Instead, it computes directly the vectors V and U without building the whole intermediate block. As a conclusion, the last method is the most stable and robust one among the analysed in this section, although less straightforward to adapt in existing BEM codes.

5 Application to low frequency electric fields induced in human tissues

The developed application has been applied to the problem involving the Laplace equation with high number of DOF in three dimensions. As an illustrative example, this section presents the calculation of the electrostatic field surrounding a grounded anatomic model of the human body (see Figure 3). In particular, current densities and electric fields induced in realistic models of the human body standing in power control room. The model shown in Figure 4 with single domain is meshed with constant triangular elements yielding nearly 50 000 and 80 000 degrees of freedom. The outcome of simulation is represented in Figure 5 which shows the current density distribution over the human body. The results obtained compare very well with similar models solved in previous work [5]. Table 1 represents

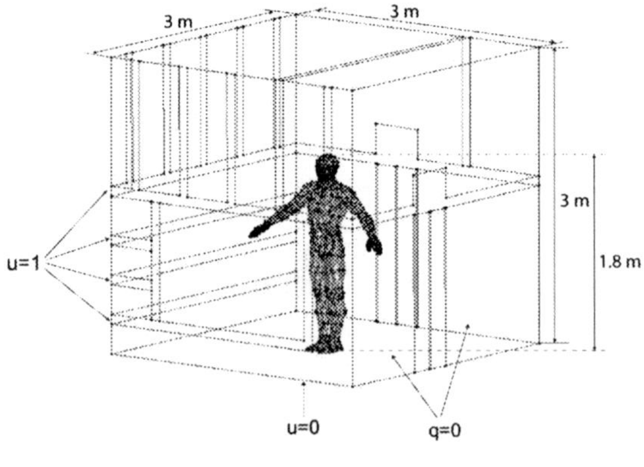

Figure 3: The BEM model under analysis: the potential $u = 1$ on each cable, the bottom has $u = 0$ as it is grounded and the normal flux $q = 0$ everywhere else as the model is isolated.

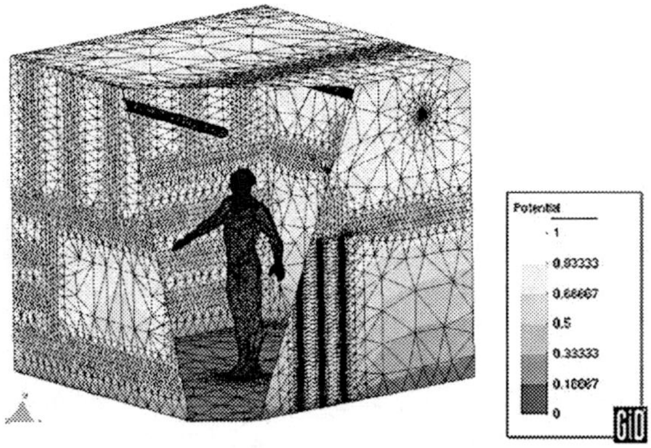

Figure 4: BEM model under analysis.

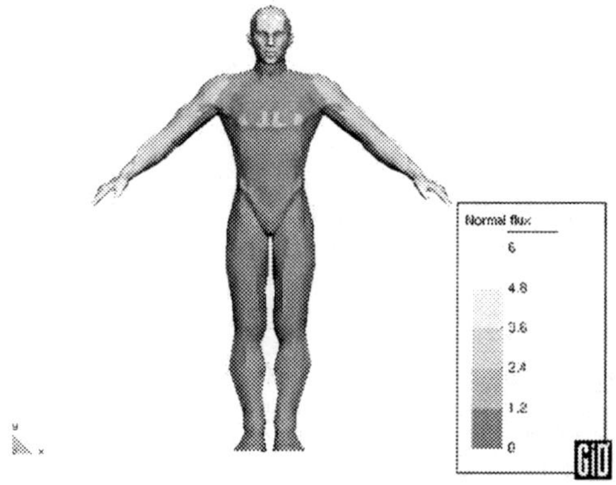

Figure 5: Normal flux distribution across the human body.

Table 1: Numerical results (Processor: Pentium iii, 1 GHz, 2 Gb RAM).

N	N_{Rk}	N_F	N_{it}	Eps	ε_{ACA}	Mem	T_p	T_s	T
47466	13124	7094	21	10^{-8}	10^{-4}	763	133	54	187
82946	91142	41012	17	10^{-8}	0.005	424	170	34	204

the numerical properties of the ACA-based solver. The following labels are used: N is the number of DOF, N_{Rk} is a number of low-rank blocks in the coefficient matrix, N_F is a number of blocks without approximation, N_{it} is a number of iterations, *Eps* in an error of iterative solver (GMRES), ε_{ACA} is an error of ACA approximation, *Mem* is a memory capacity in megabytes, T_p is a duration of the preconditioner generation in seconds, T_s is a solution time in seconds and T is a total time in seconds. The working space to allocate all necessary components can be estimated as follows: $(N * 100 * 8)/(1024 * 1024)$ Mb for temporary array, the preconditioner needs about 10% of matrix A_H space, i.e. $N * 100 * 8/10^6 + 0.1 * Mem$ Mb. So, for the biggest example, the working space is about 105 Mb.

6 Conclusions

An ACA-based solver has been implemented in order to solve linear systems of equations coming from the collocation BEM. The main advantage of the approach explained in this paper over other approaches is that it assembles both H and G BEM matrices within the same coefficient H-matrix A_H. An illustrative example explaining the assembly scheme using a very simple model of the 3D unitary cube demonstrates the approach in detail. Two different ways of using the ACA-solver within existing BEM codes have been presented. The advantages and disadvantages of both of them as well as the appropriate estimates are given in the paper. An 3D example of electrostatic model with more complex geometry has been successfully solved using the developed approach.

References

[1] M. Bebendorf and R. Grzibovski. Accelerating galerkin bem for linear elasticity using adaptive cross approximation. May 2006.

[2] S. Borm, L. Grasedyck, and W. Hackbusch. Hierarchical matrices. April 2005.

[3] C. Brebbia, J. Telles, and L. Wrobel. *Boundary Elements Techniques*. Splinger-Verlag, Berlin, Heidelberg, New York and Tokio, 1984.

[4] J. E. Flaherty T. James D., K. D. Devine. Partitioning and dynamic load balancing for the numerical solution of partial differential equations. *Num. Sol. of Part. Diff. Equations on Parallel Computers*, August 2005.

[5] A. Peratta, C. Gonzalez, and D. Poljak. Current density induced in the human body due to power distributions lines using the boundary element method. *Journal of Communications Software and Systems*, 3:11–16, 2007.

Regularization of the hypersingular integrals in 3-D problems of fracture mechanics

V. V. Zozulya
Centro de Investigación Científica de Yucatán A.C., Calle 43, No 130, Colonia Chuburná de Hidalgo, C.P. 97200, Mérida, Yucatán, México

Abstract

This article considers the hypersingular integrals, which arise when the boundary integral equation (BIE) methods are used to solve fracture mechanics problems . The methodology of hypersingular integral regularization developed in our previous publications is based on theory of distribution and Green's theorems. In the case of piecewise constant and piecewise linear approximations the hypersingular integrals are transformed into the regular contour integrals that can be easily calculated analytically or numerically.
Keywords: weakly singular, singular, hypersingular integrals, boundary integral equations, fracture mechanics.

1 Introduction

A huge amount of publications is devoted to the boundary integral equation methods (BEM) and its application science and engineering. One of the main problems arising in numerical solution of the BIE by the BEM is a calculation of the divergent integrals. Different methods have been developed for regularization of the divergent integrals [7]. The hypersingular integrals had been considered by Hadamard in the sense of finite part (*FP*) in [4]. The theory of distributions allows us to study the divergent integrals and integral operators with kernels containing different kind of singularities in the same way as the regular integrals.

Analysis of the most known methods used for treatment of the different divergent integrals has been done in our previous publications. It was shown that theory of distributions provides a unified approach for the study of the divergent integrals and integral operators with kernels containing different kind of singularities. We applied the theory of distribution approach for the first time in [8], then it was further developed in [9–11] and was applied for static and

WIT Transactions on Modelling and Simulation, Vol 47,
www.witpress.com, ISSN 1743-355X (on-line)
doi:10.2495/BE080221

dynamic problems of fracture mechanics in [17] and [18] respectively. See also monograph [3] and review articles [4, 5, 12] for details and further references.

In the present paper regularization methods developed in [9] are applied for the regularization of the hypersingular integrals that arise when the BIE methods are used for solution of the 3-D fracture mechanics problems.

2 Boundary integral equations

Let us consider a plane crack with a surface ∂V in a three-dimensional linearly elastic homogeneous isotropic space $\Re^3$. We introduce Cartesian coordinates system, with x_1 and x_2 axes in the plane of the crack, and the x_3 axis perpendicular to this plane. In [3–5] it was shown that the BIE that relate load $p_i(\mathbf{y})$ on the crack faces and crack opening $\Delta u_j(\mathbf{x})$ may be written in the following form

$$p_i(\mathbf{y}) = -\int_{\partial V} F_{ij}(\mathbf{x},\mathbf{y})\Delta u_j(\mathbf{x})\, dS. \tag{1}$$

The kernels $F_{ij}(\mathbf{x},\mathbf{y})$ in the BIE (1) may be presented in the form

$$F_{11} = \frac{\mu}{4\pi(1-\upsilon)}\left[\frac{(1-2\upsilon)}{r^3} + 3\upsilon\frac{(x_1-y_1)^2}{r^5}\right],\quad F_{12} = \frac{\mu\upsilon}{4\pi(1-\upsilon)}\frac{(x_1-y_1)(x_2-y_2)}{r^5},$$

$$F_{22} = \frac{\mu}{4\pi(1-\upsilon)}\left[\frac{(1-2\upsilon)}{r^3} + 3\upsilon\frac{(x_2-y_2)^2}{r^5}\right],\quad F_{33} = \frac{\mu}{4\pi(1-\upsilon)}\frac{1}{r^3}. \tag{2}$$

where μ and υ are elastic module and Poison ration, r is a distance between points $\mathbf{x}$ and $\mathbf{y}$

$$r = \sqrt{(x_1-y_1)^2 + (x_2-y_2)^2 + (x_3-y_3)^2}$$

3 The BEM equations

In order to transform the BIE into finite dimensional BEM equations we have to split the boundary ∂V into finite elements, which are called boundary elements (BE).

$$\partial V = \bigcup_{n=1}^{N} \partial V_n,\quad \partial V_n \cap \partial V_k = \varnothing,\quad \text{if } n \neq k.$$

On each BE we shall choose Q nodes of interpolation and shape functions $\varphi_{nq}(\mathbf{x})$. Then the vectors of displacements discontinuity and traction on the BE ∂V_n will be represented approximately in the form

$$\begin{aligned} \Delta u_i(\mathbf{x}) &\approx \sum_{q=1}^{Q} \Delta u_i^n(\mathbf{x}_q)\varphi_{nq}(\mathbf{x}),\ \mathbf{x} \in \partial V_n \\ p_i(\mathbf{x}) &\approx \sum_{q=1}^{Q} p_i^n(\mathbf{x}_q)\varphi_{nq}(\mathbf{x})\ \mathbf{x} \in \partial V_n \end{aligned} \tag{3}$$

and on the hold boundary ∂V in the form

$$\Delta u_i(\mathbf{x}) \approx \sum_{n=1}^{N}\sum_{q=1}^{Q} \Delta u_i^n(\mathbf{x}_q)\varphi_{nq}(\mathbf{x}), \quad \mathbf{x} \in \bigcup_{n=1}^{N} \partial V_n,$$
$$p_i(\mathbf{x}) \approx \sum_{n=1}^{N}\sum_{q=1}^{Q} p_i^n(\mathbf{x}_q)\varphi_{nq}(\mathbf{x}), \quad \mathbf{x} \in \bigcup_{n=1}^{N} \partial V_n \tag{4}$$

Substituting expressions (4) in (1) gives us the BE equations which relate the vectors of displacements discontinuity and traction on the crack surface in the form

$$p_i^m(\mathbf{y}_r) = -\sum_{n=1}^{N}\sum_{q=1}^{Q} F_{ji}^n(\mathbf{y}_r, \mathbf{x}_q)\Delta u_j^n(\mathbf{x}_q) \tag{5}$$

where

$$F_{ji}^n(\mathbf{y}_r, \mathbf{x}_q) = \int_{\partial V_n} F_{ji}(\mathbf{y}_r, \mathbf{x})\varphi_{nq}(\mathbf{x})dS. \tag{6}$$

More detailed information on transition from the BIE to the BEM equations can be found in [1,2].

4 Piecewise constant approximation

In the application of in the BEM, it is necessary to calculate the divergent integrals over any triangular, rectangular or polygonal elements. The piecewise constant approximation is the simplest one. Interpolation functions in this case do not depend on the BE form and dimension of the domain. They have the form

$$\varphi_{nq}(\mathbf{x}) = \begin{cases} 1 & \forall \mathbf{x} \in \partial V_n, \\ 0 & \forall \mathbf{x} \notin \partial V_n. \end{cases} \tag{7}$$

In order to simplify situation we transform origin of the global system of coordinates at the nodal point, where $\mathbf{y}^0 = 0$. Regular representations for the hypersingular integrals in (6) have the form

$$J_3^{0,0} = F.P.\int_{S_n} \frac{dS}{r^3} = -\int_{\partial S_n} \frac{r_n}{r^3} dl, \quad J_5^{2,0} = F.P.\int_{S_n} \frac{x_1^2}{r^5} dS = \int_{\partial S_n} \left(\frac{r_n}{3r^3} - \frac{x_2^2 r_n}{r^5} - \frac{2x_1 n_1}{3r^3} \right) dl,$$
$$J_5^{0,2} = F.P.\int_{S_n} \frac{x_2^2}{r^5} dS = \int_{\partial S_n} \left(\frac{r_n}{3r^3} - \frac{x_1^2 r_n}{r^5} - \frac{2x_2 n_2}{3r^3} \right) dl, \tag{8}$$
$$J_5^{1,1} = F.P.\int_{S_n} \frac{x_1 x_2}{r^5} dS = \int_{\partial S_n} \left(\frac{x_1 x_2 r_n}{r^5} - \frac{r_+}{3r^3} \right) dl$$

where $r_n = x_\alpha n_\alpha$, $r_+ = x_1 n_2 + x_2 n_1$, $r_- = x_2 n_1 - x_1 n_2$. See [10, 11, 13] for details.

Let us consider the contour ∂V_n as a polygon with K vertexes us it is shown on fig. 1.

In this case

$$F_{ji}^n(\mathbf{y}_r, \mathbf{x}_q) = \sum_{k=1}^{K} \int_{l_k} F_{ji}(\mathbf{y}_r, \mathbf{x})dl. \tag{9}$$

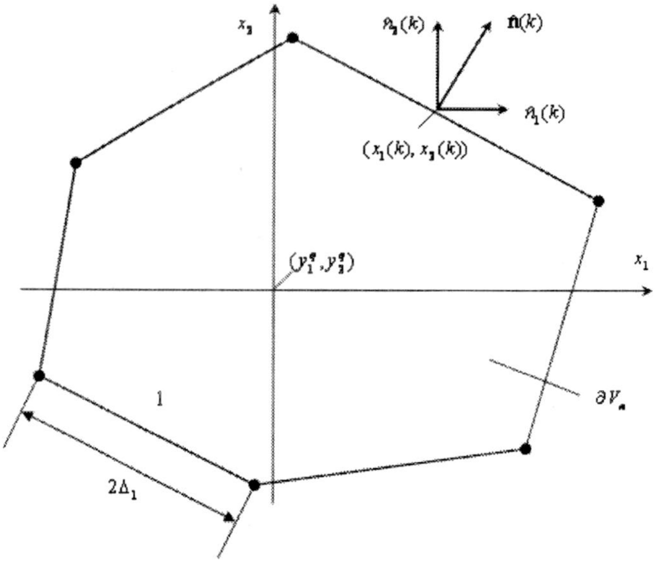

Figure 1.

Coordinates of an arbitrary point on the contour ∂V_n may be represented in the form

$$x_1(\xi) = x_1(k) - \xi\Delta_k \hat{n}_2 \text{ and } x_2(\xi) = x_2(k) + \xi\Delta_k \hat{n}_1 \qquad (10)$$

where $x_1(k)$ and $x_2(k)$ are the coordinates of the middle of k-th side of the contour, $\hat{\mathbf{n}}(\hat{n}_1, \hat{n}_2)$ is a unit vector normal to the contour and $\xi \in [-1,1]$ is a parameter of integration along the k-th side, $2\Delta_k$ is the length of a k-th side.

These are some useful notations which will be used bellow

$$r(\xi) = \sqrt{\Delta_k^2\xi^2 + 2\xi\Delta_k r_-(k) + r^2(k)}\ , r(k) = \sqrt{x_1^2(k) + x_2^2(k)}\ , r_n(k) = x_\alpha(k)\hat{n}_\alpha(k),$$
$$r_\tau(k) = x_1(k)\hat{n}_1(k) - x_2(k)\hat{n}_2(k),\ r_+(k) = x_1(k)\hat{n}_2(k) + x_2(k)\hat{n}_1(k), \qquad (11)$$

Using these notations the integrals in (8) may be represented in a convenient for the calculation form

$$J_{0,3}^{0,0}(k) = -r_n(k)I_{3,0},$$

$$J_{0,5}^{2,0}(k) = \frac{1}{3}r_n(k)I_{3,0} - r_n(k)\left(x_2^2(k)I_{5,0} - 2\hat{n}_1(k)x_2(k)I_{5,1} - \hat{n}_1^2(k)I_{5,2}\right) - \\ -\frac{2}{3}\left(\hat{n}_1(k)\left(x_2(k)I_{3,0} - \hat{n}_2(k)I_{3,1}\right)\right)$$

$$J_{0,5}^{0,2}(k) = \frac{1}{3}r_n(k)I_{3,0} - r_n(k)\left(x_1^2(k)I_{5,0} + 2\hat{n}_2(k)x_1(k)I_{5,1} - \hat{n}_2^2(k)I_{5,2}\right) - \\ -\frac{2}{3}\left(\hat{n}_2(k)\left(x_1(k)I_{3,0} + \hat{n}_2(k)I_{3,1}\right)\right) \qquad (12)$$

$$J_{0,5}^{1,1}(k) = r_n(k)\left(x_1(k)x_2(k)I_{5,0} + r_\tau(k)I_{5,1} - \hat{n}_1(k)\hat{n}_2(k)I_{5,1}\right) - \\ -\frac{1}{3}(r_+(k)I_{5,0} + (\hat{n}_1^2(k) - \hat{n}_2^2(k))I_{5,1}$$

Integrals in (12) may be calculated analytically

$$I_{1,0} = \Delta_k \int_{-1}^{1} \frac{1}{r(\xi)} d\xi = \ln\left|r_\tau(k) + \Delta_k \xi + r(\xi)\right|\Big|_{-1}^{1},$$

$$I_{1,1} = (\Delta_k)^2 \int_{-1}^{1} \frac{\xi}{r(t)} d\xi = r(\xi)\Big|_{-1}^{1} - r_\tau(k)I_{1,0},$$

$$I_{3,0} = \Delta_k \int_{-1}^{1} \frac{1}{r(\xi)^3} d\xi = \frac{1}{r^2(k) - r_\tau^2(k)} \frac{\Delta_k \xi + r_\tau(k)}{r(\xi)}\Bigg|_{-1}^{1},$$

$$I_{3,1} = (\Delta_k)^2 \int_{-1}^{1} \frac{\xi}{r(\xi)^3} d\xi = -\frac{1}{r^2(k) - r_\tau^2(k)} \frac{r_\tau(k)\Delta_k \xi + r^2(k)}{r(\xi)}\Bigg|_{-1}^{1},$$

$$I_{5,0} = \Delta_k \int_{-1}^{1} \frac{1}{r(\xi)^5} d\xi = \frac{1}{r^2(k) - r_\tau^2(k)} I_{3,0} + \frac{1}{2\left(r^2(k) - r_\tau^2(k)\right)} \frac{\Delta_k \xi + r_\tau(k)}{r(\xi)^3}\Bigg|_{-1}^{1}, \quad (13)$$

$$I_{5,1} = (\Delta_k)^2 \int_{-1}^{1} \frac{\xi}{r(\xi)^5} d\xi = r_\tau(k)I_{5,0} - \frac{1}{3r(\xi)^3}\Bigg|_{-1}^{1},$$

$$I_{5,2} = (\Delta_k)^3 \int_{-1}^{1} \frac{\xi^2}{r(\xi)^5} d\xi = \frac{(\Delta_k \xi + r_\tau(k))^3}{3(r^2 - r_\tau^2)r(\xi)^3}\Bigg|_{-1}^{1} - 2r_\tau(k)I_{5,1} - r^2(k)I_{5,0}.$$

Then fundamental solutions (9) may be represented in the form

$$F_{11}''\left(\mathbf{y}_r, \mathbf{x}_q\right) = \frac{\mu}{4\pi(1-\upsilon)}\left[(1-2\upsilon)\sum_{k=1}^{K} J_{0,3}^{0,0}(k) + 3\nu\sum_{k=1}^{K} J_{0,5}^{0,2}(k)\right],$$

$$F_{22}''\left(\mathbf{y}_r, \mathbf{x}_q\right) = \frac{\mu}{4\pi(1-\upsilon)}\left[(1-2\upsilon)\sum_{k=1}^{K} J_{0,3}^{0,0}(k) + 3\nu\sum_{k=1}^{K} J_{0,5}^{2,0}(k)\right], \quad (14)$$

$$F_{33}''\left(\mathbf{y}_r, \mathbf{x}_q\right) = -\frac{\mu}{4\pi(1-\upsilon)}\sum_{k=1}^{K} J_{0,3}^{0,0}(k), \quad F_{12}''\left(\mathbf{y}_r, \mathbf{x}_q\right) = \frac{\mu\upsilon}{4\pi(1-\upsilon)}\sum_{k=1}^{K} J_{0,5}^{1,1}(k)$$

It is important to mention that here all calculations can be done analytically, no numerical integration is needed.

5 Piecewise linear approximation

Let us consider rectangle BE that is shown on fig. 2.
The quadrilateral BE is defined by it's angular nodes and its shape functions are

$$\varphi_1 = 1/4\left(1-\xi_1\right)\left(1-\xi_2\right) \qquad \varphi_2 = 1/4\left(1+\xi_1\right)\left(1-\xi_2\right)$$
$$\varphi_3 = 1/4\left(1+\xi_1\right)\left(1+\xi_2\right) \qquad \varphi_4 = 1/4\left(1-\xi_1\right)\left(1+\xi_2\right) \quad (15)$$

According to (5) and (6) in this case we have to calculate

$$F_{ji}^{n}(\mathbf{y}_r,\mathbf{x}_q)=\sum_{q=1}^{4}\int_{\partial V_n}F_{ji}(\mathbf{y}_r,\mathbf{x})\varphi_q(\mathbf{x})dS=\sum_{q=1}^{4}\sum_{k=1}^{4}\int_{l_k}F_{ji}(\mathbf{y}_r,\mathbf{x})\varphi_q(\mathbf{x})dl. \qquad (16)$$

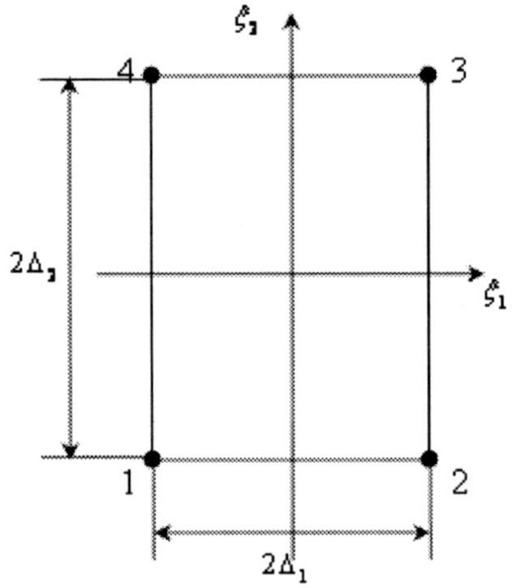

Figure 2.

Regular representations for the hypersingular integrals in (6) in this case have the form

$$J_{q,3}^{0,0}=F.P.\int_{S_n}\frac{\varphi_q(\xi)}{r^3}dS=-\int_{\partial S_n}\left(\varphi_q(\xi)\frac{r_n}{r^3}+\frac{1}{r}\partial_n\varphi_q(\xi)\right)dl\,, \qquad (17)$$

$$J_{q,5}^{2,0}=F.P.\int_{S_n}\varphi_q(\xi)\frac{x_1^2}{r^5}dS=\int_{\partial S_n}\left(\varphi_q(\xi)\left(\frac{r_n}{3r^3}-\frac{x_2^2r_n}{r^5}-\frac{2x_1n_1}{3r^3}\right)-\left(\frac{1}{r}+\frac{x_1^2}{r^3}\right)\partial_n\varphi_q(\xi)\right)dl\,,$$

$$J_{q,5}^{0,2}=F.P.\int_{S_n}\varphi_q(\xi)\frac{x_2^2}{r^5}dS=\int_{\partial S_n}\left(\varphi_q(\xi)\left(\frac{r_n}{3r^3}-\frac{x_1^2r_n}{r^5}-\frac{2x_2n_2}{3r^3}\right)-\left(\frac{1}{r}+\frac{x_2^2}{r^3}\right)\partial_n\varphi_q(\xi)\right)dl\,,$$

$$J_{q,5}^{1,1}=F.P.\int_{S_n}\varphi_q(\xi)\frac{x_1x_2}{r^5}dS=\int_{\partial S_n}\left(\varphi_q(\xi)\left(\frac{x_1x_2r_n}{r^5}-\frac{r_+}{3r^3}\right)-\frac{x_1x_2}{r^3}\partial_n\varphi_q(\xi)\right)dl$$

In order to simplify situation we transform origin of the global system of coordinates at the middle point of re. The coordinate axes x_1 and x_2 are located in the plane of the BE and coincide with the local ones ξ_1 and ξ_2, while the axis x_3 is perpendicular to that plane. Coordinates of the nodal points are

$$(x_1^1=-\Delta_1,x_2^1=-\Delta_2),\ (x_1^2=\Delta_1,x_2^2=-\Delta_2),$$
$$(x_1^3=\Delta_1,x_2^3=\Delta_2),\ (x_1^4=\Delta_1,x_2^4=-\Delta_2)$$

These are some useful notations which will be used bellow

$$x_1 = \Delta_1(1+\xi_1),\ x_2 = \Delta_2(1+\xi_2),\ dS = \Delta_1\Delta_2 d\xi_1 d\xi_2,\ dl = \sqrt{dx_1^2 + dx_2^2}$$

$$r(\xi, \mathbf{y}^q) = \sqrt{(x_1 - y_1^q)^2 + (x_2 - y_2^q)^2} = \sqrt{(\Delta_1(1+\xi_1) - y_1^q)^2 + (\Delta_2(1+\xi_2) - y_2^q)^2},$$

$$r_n = (x_\alpha - y_\alpha^q)n_\alpha,\ r_+ = (x_1 - y_1^q)n_2 + (x_2 - y_2^q)n_1,\ r_- = (x_2 - y_2^q)n_1 - (x_1 - y_1^q)n_2$$

$$\partial_n\varphi_1(\xi) = \frac{1}{4}(r_+(\xi) - \hat{n}_1 - \hat{n}_2),\ \partial_n\varphi_2(\xi) = \frac{1}{4}(-r_+(\xi) + \hat{n}_1 - \hat{n}_2), \tag{18}$$

$$\partial_n\varphi_3(\xi) = \frac{1}{4}(r_+(\xi) + \hat{n}_1 + \hat{n}_2),\ \partial_n\varphi_4(\xi) = \frac{1}{4}(-r_+(\xi) - \hat{n}_1 + \hat{n}_2).$$

Now we will calculate integrals (17) that are included in representation (16) for fundamental solutions. Let point $\mathbf{y}^0$ is located at the left low corner of the rectangle (point 1). Its coordinates are $\mathbf{y}^0 = (-\Delta_1, -\Delta_2)$. Other point can be considered in the same way. In this case a situation is not symmetrical we have to do calculations side by side. Order of summation in (16) can be changed, therefore we first will calculate integrals in (17) for every shape function and then their sums

$$J_{I,n}^{l,m}(k) = \sum_{q=1}^{4} J_{q,n}^{l,m} \tag{19}$$

Side 1-2. In this case $n_1 = 0$, $n_2 = -1$, $\xi_2 = -1$. The main parameters defined by (18) are

$$x_1 = \Delta_1(1+\xi_1),\ x_2 = -\Delta_2,\ dl = \Delta_1 d\xi_1,\ r(\xi_1) = \Delta_1(1+\xi_1),\ r_+ = -\Delta_1(1+\xi_1),$$

$$r_- = \Delta_1(1+\xi_1),\ r_n = 0,\ \varphi_1 = 1/2(1-\xi_1),\ \varphi_2 = 1/2(1+\xi_1),\ \varphi_3 = 0,\ \varphi_4 = 0,$$

$$\partial_n\varphi_1(\xi) = 1/4(1-\xi_1), \partial_n\varphi_2(\xi) = 1/4(1+\xi_1), \partial_n\varphi_3(\xi) = -1/4(1+\xi_1),$$

$$\partial_n\varphi_4(\xi) = -1/4(1-\xi_1)$$

The sums (19) in this case are

$$J_{I,3}^{0,0}(1) = \sum_{q=1}^{4} J_{q,3}^{0,0} = 0,\ J_{I,5}^{2,0}(1) = \sum_{q=1}^{4} J_{q,5}^{2,0} = 0,\ J_{I,5}^{0,2}(1) = \sum_{q=1}^{4} J_{q,5}^{0,2} = 0,$$

$$J_{I,5}^{1,1}(1) = \sum_{q=1}^{4} J_{q,5}^{1,1} = -\frac{1}{6\Delta_1}$$

Side 2-3. In this case $n_1 = 1$, $n_2 = 0$, $\xi_1 = 1$, The main parameters defined by (18) are

$$x_1 = \Delta_1, x_2 = \Delta_2(1+\xi_2),\ dl = \Delta_2 d\xi_2,\ r(\xi_2) = \sqrt{4\Delta_1^2 + \Delta_2^2(1+\xi_2)^2},\ r_n = 2\Delta_1,$$

$$r_+ = \Delta_2(1+\xi_2),\ r_- = \Delta_2(1+\xi_2),\ \varphi_1 = 0,\ \varphi_2 = 1/2(1-\xi_2),\ \varphi_3 = 1/2(1+\xi_2),$$

$$\varphi_4 = 0,\ \partial_n\varphi_1(\xi) = -1/4(1-\xi_2),\ \partial_n\varphi_2(\xi) = 1/4(1-\xi_2),$$

$$\partial_n\varphi_3(\xi) = 1/4(1+\xi_2), \partial_n\varphi_4(\xi) = -1/4(1+\xi_2)$$

The sums (19) in this case are

$$J_3^{0,0}(2)=\sum_{q=1}^{4}\Big(\varepsilon_q^2\Big(J_{0,1}^{0,0}(2)+\zeta_q^2 J_{0,3}^{0,0}\Big|_1(2)\Big)+$$
$$+\frac{1}{2}(r_n^q(2)I_{1,0}(2)+\xi_1^q\xi_2^q(\hat{n}_1(2)+\hat{n}_2(2))I_{1,0}\Big|_1(2))\Big)$$

$$J_5^{2,0}(2)=\frac{1}{2}\sum_{q=1}^{4}\Big(\varepsilon_q^2\Big(J_{0,5}^{2,0}(2)+\zeta_q^2 J_{0,5}^{2,0}\Big|_1(2)\Big)-$$
$$-\frac{1}{2}\Big(r_n^q(2)J_3^{2,0}(2)+\xi_1^q\xi_2^q(\hat{n}_1(2)+\hat{n}_2(2))x_2(2)J_3^{2,0}\Big|_1(2)\Big)\Big)$$

$$J_{1,5}^{0,2}(2)=\frac{1}{6}\sum_{q=1}^{4}\Big(\varepsilon_q^2\Big(J_{0,5}^{0,2}(2)+\zeta_q^2 J_{0,5}^{0,2}\Big|_1(2)\Big)-$$
$$-\frac{1}{2}\Big(r_n^q(2)J_3^{0,2}(2)+\xi_1^q\xi_2^q(\hat{n}_1(2)+\hat{n}_2(2))x_2(2)J_3^{0,2}\Big|_1(2)\Big)\Big)$$

$$J_{1,5}^{1,1}(2)=\frac{1}{6}\sum_{q=1}^{4}\Big(\varepsilon_q^2\Big(J_{0,5}^{1,1}(2)+\zeta_q^2 J_{0,5}^{1,1}\Big|_1(2)\Big)-$$
$$-\frac{1}{2}\Big(r_n^q(2)J_3^{1,1}(2)+\xi_1^q\xi_2^q(\hat{n}_1(2)+\hat{n}_2(2))x_2(2)J_3^{1,1}\Big|_1(2)\Big)\Big)$$

Here $\varepsilon_q^k=\begin{cases}1 & \text{for}\quad k=q\,\&\,k=q+1\\ 0 & \text{otherwise}\end{cases}$, $I_{m,l-1}\Big|_1=(\Delta_k)^l\int_{-1}^{1}\frac{\xi^l}{r^m(\xi)}d\xi$,

$$J_3^{2,0}(k)=x_1^2(k)I_{3,0}+2\hat{n}_2(k)x_1(k)I_{3,1}-\hat{n}_2^2(k)I_{3,2}+I_{1,0}(k),$$
$$J_3^{0,2}(k)=x_2^2(k)I_{3,0}-2\hat{n}_1(k)x_2(k)I_{3,1}-\hat{n}_1^2(k)I_{3,2}+I_{1,0}(k).$$

Side 3-4. In this case $n_1=0$, $n_2=1$, $\xi_2=1$. The main parameters defined by (18) are

$$x_1=\Delta_1(1+\xi_1),\ x_2=\Delta_2,\ dl=\Delta_1 d\xi_1,\ r(\xi_1)=\sqrt{4\Delta_2^2+\Delta_1^2(1+\xi_1)^2},\ r_n=2\Delta_2,$$
$$r_+=\Delta_1(1+\xi_1),\ r_-=-\Delta_1(1+\xi_1),\ \varphi_1=0,\ \varphi_2=0,\ \varphi_3=1/2(1+\xi_1),$$
$$\varphi_4=1/2(1-\xi_1),\ \partial_n\varphi_1(\xi)=-1/4(1-\xi_1),\ \partial_n\varphi_2(\xi)=-1/4(1+\xi_1),$$
$$\partial_n\varphi_3(\xi)=1/4(1+\xi_1),\ \partial_n\varphi_4(\xi)=1/4(1-\xi_1)$$

The sums (19) in this case are

$$J_3^{0,0}(3)=\sum_{q=1}^{4}\Big(\varepsilon_q^3\Big(J_{0,1}^{0,0}(2)+\zeta_q^3 J_{0,3}^{0,0}\Big|_1(3)\Big)+$$
$$+\frac{1}{2}(r_n^q(3)I_{1,0}(3)+\xi_1^q\xi_2^q(\hat{n}_1(3)+\hat{n}_2(3))I_{1,0}\Big|_1(3))\Big)$$

$$J_5^{2,0}(3)=\frac{1}{2}\sum_{q=1}^{4}\Big(\varepsilon_q^3\Big(J_{0,5}^{2,0}(3)+\zeta_q^3 J_{0,5}^{2,0}\Big|_1(3)\Big)-$$

$$-\frac{1}{2}\left(r_n^q(3)J_3^{2,0}(3)+\xi_1^q\xi_2^q(\hat{n}_1(3)+\hat{n}_2(3))x_2(3)J_3^{2,0}\Big|_1(3)\right)\Bigg)$$

$$J_{1,5}^{0,2}(3)=\frac{1}{6}\sum_{q=1}^{4}\left(\varepsilon_q^3\left(J_{0,5}^{0,2}(3)+\zeta_q^3 J_{0,5}^{0,2}\Big|_1(3)\right)-\right.$$

$$\left.-\frac{1}{2}\left(r_n^q(3)J_3^{0,2}(3)+\xi_1^q\xi_2^q(\hat{n}_1(3)+\hat{n}_2(3))x_2(3)J_3^{0,2}\Big|_1(3)\right)\right)$$

$$J_{1,5}^{1,1}(3)=\frac{1}{6}\sum_{q=1}^{4}\left(\varepsilon_q^3\left(J_{0,5}^{1,1}(3)+\zeta_q^2 J_{0,5}^{1,1}\Big|_1(3)\right)-\right.$$

$$\left.-\frac{1}{2}\left(r_n^q(3)J_3^{1,1}(3)+\xi_1^q\xi_2^q(\hat{n}_1(3)+\hat{n}_2(3))x_2(3)J_3^{1,1}\Big|_1(3)\right)\right)$$

Side 4-1. In this case $n_1=-1$, $n_2=0$, $\xi_1=-1$. The main parameters defined by (18) are

$$x_1=-\Delta_1,\ x_2=\Delta_2(1+\xi_2),\ dl=\Delta_2 d\xi_2\ r(\xi_2)=\Delta_2(1+\xi_2),\ r_n=0,$$

$$r_+=-\Delta_2(1+\xi_2),\ r_-=0,\ \varphi_1=1/2(1-\xi_2),\ \varphi_2=0,\ \varphi_3=0,\ \varphi_4=1/2(1+\xi_2),$$

$$\partial_n\varphi_1(\xi)=1/4(1-\xi_2),\ \partial_n\varphi_2(\xi)=-1/4(1-\xi_2),\ \partial_n\varphi_3(\xi)=-1/4(1+\xi_2),$$

$$\partial_n\varphi_4(\xi)=1/4(1+\xi_2)$$

The sums (19) in this case are

$$J_{I,3}^{0,0}(4)=\sum_{q=1}^{4}J_{q,3}^{0,0}=0,\ J_{I,5}^{2,0}(4)=\sum_{q=1}^{4}J_{q,5}^{2,0}=0,\ J_{I,5}^{0,2}(4)=\sum_{q=1}^{4}J_{q,5}^{0,2}=0,$$

$$J_{I,5}^{1,1}(4)=\sum_{q=1}^{4}J_{q,5}^{1,1}=\frac{1}{6\Delta_2}$$

Substituting all obtained for each side equations in (16) finally we have

$$F_{11}^n\left(\mathbf{y}_r,\mathbf{x}_q\right)=\frac{\mu}{4\pi(1-\upsilon)}\left[(1-2\upsilon)\sum_{k=1}^{K}J_{I,3}^{0,0}(k)+3v\sum_{k=1}^{K}J_{I,5}^{0,2}(k)\right],$$

$$F_{22}^n\left(\mathbf{y}_r,\mathbf{x}_q\right)=\frac{\mu}{4\pi(1-\upsilon)}\left[(1-2\upsilon)\sum_{k=1}^{K}J_{I,3}^{0,0}(k)+3v\sum_{k=1}^{K}J_{I,5}^{2,0}(k)\right],\qquad(20)$$

$$F_{33}^n\left(\mathbf{y}_r,\mathbf{x}_q\right)=-\frac{\mu}{4\pi(1-\upsilon)}\sum_{k=1}^{K}J_{I,3}^{0,0}(k),\ F_{12}^n\left(\mathbf{y}_r,\mathbf{x}_q\right)=\frac{\mu\upsilon}{4\pi(1-\upsilon)}\sum_{k=1}^{K}J_{I,5}^{1,1}(k)$$

It is important to mention that here all calculations can be done analytically, no numerical integration is needed.

References

[1] Brebbia C.A., Dominguez J. *Boundary elements. An introductory course*, WIT Press, Southampton, 1998

[2] Brebbia C.A., Telles J.C.F. and Wrobel L.C. *Boundary element techniques. Theory and applications in engineering.* Springer-Verlag, Berlin. 1984

[3] Guz A.N. and Zozulya V.V., *Brittle fracture of constructive materials under dynamic loading*, Kiev, Naukova Dumka, 1993, (in Russian).
[4] Guz A.N. and Zozulya V.V. Fracture dynamics with allowance for a crack edges contact interaction, *International Journal of Nonlinear Sciences and Numerical Simulation,* **2**(3), pp. 173–233, 2001.
[5] Guz A.N. and Zozulya V.V. Elastodynamic unilateral contact problems with friction for bodies with cracks, *International Applied Mechanics,* **38**(8), pp. 3–45, 2002.
[6] Hadamard J., *Le probleme de cauchy et les eguations aux devivees partielles lineaires hyperboliques,* Herman, Paris, 1932, [English translation, Dover, New York, 1952], [Russian translation: Nauka, Moscow, 1978]
[7] Tanaka M., Sladek V. and Sladek J., Regularization techniques applied to boundary element methods. *Applied Mechanics Review*, **47**, pp. 457–499, 1994.
[8] Zozulya V.V., Integrals of Hadamard type in dynamic problem of the crack theory. *Doklady Academii Nauk. UkrSSR, Ser. A. Physical Mathematical & Technical Sciences*, **2**, pp.19–22, 1991, (in Russian).
[9] Zozulya V.V. Regularization of the divergent integrals. I. General consideration. *Electronic Journal of Boundary Elements,* **4**(2), pp. 49–57, 2006.
[10] Zozulya V.V. Regularization of the divergent integrals. II. Application in Fracture Mechanics. *Electronic Journal of Boundary Elements,* **4**(2), pp. 58–56, 2006.
[11] Zozulya V.V. and Gonzalez-Chi P.I. Weakly singular, singular and hypersingular integrals in elasticity and fracture mechanics, *Journal of the Chinese Institute of Engineers,* **22**(6), pp. 763–775, 1999.
[12] Zozulya V.V. and Gonzalez-Chi P.I. Dynamic fracture mechanics with crack edges contact interaction, *Engineering Analysis with Boundary Elements,* **24**(9), pp. 643–659, 2000.
[13] Zozulya V.V. and Lukin A.N. Solution of three-dimensional problems of fracture mechanics by the method of integral boundary equations. *International Applied Mechanics,* **34**(6), pp. 544–551, 1998.
[14] Zozulya V.V. and Men'shikov V.A. Solution of tree dimensional problems of the dynamic theory of elasticity for bodies with cracks using hypersingular integrals, *International Applied Mechanics*, **36**(1), pp. 74–81, 2000.

Efficient elasto-plastic analysis via an adaptive finite element-boundary element coupling method

W. Elleithy & U. Langer
Institute of Computational Mathematics,
Johannes Kepler University Linz, Altenberger Str. 69, A-4040 Linz, Austria

Abstract

This paper presents an adaptive FEM-BEM coupling method for elasto-plastic analysis. The proposed method is valid for both two- and three-dimensional applications. The method takes care of the evolution of the elastic and plastic regions and avoids some limitations of the standard FEM-BEM coupling approaches. It estimates the FEM and BEM sub-domains and automatically generates/adapts the FEM and BEM meshes/sub-domains, according to the state of computation. The method eliminates the cumbersome trial and error process in the identification of the FEM and BEM sub-domains in the standard FEM-BEM coupling approaches. An example application confirms the effectiveness of the proposed method.
Keywords: FEM, BEM, elasto-plasticity, adaptive coupling.

1 Introduction

There exist many application contexts where coupling of the finite element method (FEM) and the boundary element method (BEM) is, in principle, very attractive. Examples include, but not limited to, elasto-plastic applications with limited spread of plastic deformations. The FEM is utilized where the plastic material behaviour is expected to develop. The remaining bounded/unbounded linear elastic regions are best approximated by the BEM.

A crucial aspect of the existing (standard) FEM-BEM coupling approaches is that they require the user/analyst to predefine and manually localize the FEM and

WIT Transactions on Modelling and Simulation, Vol 47, © 2008 WIT Press
www.witpress.com, ISSN 1743-355X (on-line)
doi:10.2495/BE080231

BEM sub-domains (prior to analysis). In an elasto-plastic analysis, it is difficult or impossible to predict regions where plasticity occurs. If the FEM sub-domain is not predefined in order to enclose the evolutional plastic regions, significant errors will be introduced to the conducted FEM-BEM coupling analysis. On the other hand, if the FEM sub-domain is notably over-estimated, the excessive number of degrees of freedom (in order to model the linear elastic part) will significantly increase the computational demand.

In pure BEM elasto-plastic analysis, the conventional numerical implementation requires the domain to be discretized into cells (predefined by the user/analyst). In this case the BEM loses its main advantage. Astrinidis et al. [1] presented adaptive discretization schemes that are based on a stress smoothing error criterion in the case of 2D elastic analysis, and on a total strain smoothing error criterion in the case of 2D elasto-plasticity. Maischak and Stephan [2] showed convergence for the boundary element approximation, obtained by the *hp*-version, for elastic contact problems, and derived a posteriori error estimates together with error indicators for adaptive *hp*-algorithms. Rebeiro et al. [3] developed a pure BEM approach to automatically generate the internal cells. Their approach considers cases when plasticity starts from the boundary. The stresses at the boundary nodes are computed and the internal cells are generated surrounding regions where plasticity is detected. The process is iterated until the detected plastic regions are fully discretized.

Brink et al. [4] investigated the coupling of mixed finite elements and Galerkin boundary elements in linear elasticity, taking into account adaptive mesh refinement based on a posteriori error estimators. Carstensen et al. [5] presented an *h*-adaptive FEM-BEM coupling approach (mesh refinement of the boundary elements and the finite elements) for the solution of visco-plastic and elasto-plastic interface problems. Mund and Stephan [6] derived a posteriori error estimates for nonlinear coupled FEM-BEM equations by using hierarchical basis techniques. They presented an approach for adaptive error control, which allows independent refinements of the finite and boundary elements.

Doherty and Deeks [7] developed an adaptive approach for analyzing 2D elasto-plastic unbounded media by coupling the FEM with the scaled boundary finite element method. The analysis begins with an "initial" finite element mesh that tightly encloses the load-medium interface, whereas the remainder of the problem is modelled using the semi-analytical scaled boundary finite element method. Load increments are applied, and if plasticity is detected in the outer band of finite elements, an additional band is added around the perimeter of the existing mesh. The scaled boundary finite element sub-domain is stepped out accordingly. However, this approach requires, in general, a preliminary knowledge of the parts of the domain that are likely to yield. Moreover, it requires additional iterations when plasticity is detected in the outer band of finite elements in order to accurately determine the computational sub-domains. This may end up in a very time-consuming process.

Elleithy and Langer [8] and Elleithy [9] presented an adaptive FEM-BEM coupling method for 2D elasto-plastic analysis. The method proposes the use of simple, fast post-calculations, based on energetic methods, and follows a simple hypothetical elastic boundary element computation in order to give fast and helpful estimation of the FEM and BEM sub-domains.

In this paper we present an alternative to the adaptive coupling method presented in [8,9]. The proposed method is valid for both two- and three-dimensional applications. The adaptive coupling method improves the estimation of regions where plastic material behaviour is going to develop. An outline of the paper is as follows. Section 2 briefly summarizes the conventional FEM-BEM coupling equations adopted in this investigation and the adaptive method of [8,9]. In the sequence, Section 3 presents the proposed adaptive FEM-BEM coupling method for elasto-plastic analysis. In Section 4, we present a numerical example that highlights the effectiveness of the adaptive coupling method.

2 Preliminaries

In this section, the conventional (direct) FEM-BEM coupling equations adopted in this investigation (without loss of generality) and the adaptive method of [8,9] are briefly described.

Elasto-plastic problems with limited spread of plastic strains lend themselves to a coupled approach. The FEM is utilized in regions where plastic material behaviour is expected to develop, whereas the complementary bounded/unbounded linear elastic region is approximated using the symmetric Galerkin BEM. The domain of the original problem Ω (with known boundary conditions specified on the entire boundary $\Gamma = \Gamma_N \cup \Gamma_D$) is decomposed into two sub-domains, namely ${}_F\Omega$ and ${}_B\Omega$, with the FEM-BEM coupling interface Γ_C. The coupled FEM-BEM equations, in incremental form, are solved at each iteration using a tangent FEM stiffness matrix K_T. The coupled FEM-BEM equations for a typical iteration may then be reformulated in the symmetric form

$$\begin{bmatrix} {}_F K_{T\,FF} & {}_F K_{T\,FC} & \\ {}_F K_{T\,CF} & {}_F K_{T\,CC} + {}_B K_{CC} & {}_B K_{CB} \\ & {}_B K_{BC} & {}_B K_{BB} \end{bmatrix} \begin{bmatrix} \Delta_F \underline{u}_F \\ \Delta \underline{u}_C \\ \Delta_B \underline{u}_B \end{bmatrix} = - \begin{bmatrix} \Delta_F \underline{\psi}_F \\ \Delta \underline{\psi}_C \\ \Delta_B \underline{\psi}_B \end{bmatrix}, \tag{1}$$

where $\underline{u}$ and $\underline{\psi}$ are the displacements and the residual force vectors, respectively. The subscripts $(\,)_F$ and $(\,)_B$ indicate the displacement vectors (force vectors) not associated with the FEM and BEM sub-domains interface, respectively. The subscript $(\,)_C$ indicates those associated with the interface Γ_C.

Ref. [8,9] presented an adaptive FEM-BEM coupling method for solving problems in elasto-plasticity. The adaptive coupling method follows a linear hypothetical elastic computation at levels of loading specified by the user. The hypothetical elastic state of stresses is checked against yielding with a pseudo value of the material yield strength. An estimate of the regions sensible for FEM discretization is then derived. The FEM and BEM meshes are automatically generated. A coupled FEM-BEM stress analysis involving elasto-plastic deformations is then conducted. In order to determine the pseudo value of the material yield strength, an energy balance between the hypothetical elastic and elasto-plastic calculations was assumed in [8,9]

$$U_{\text{hyp elastic}} := \Big(\int_{\Omega} \sigma_{ij}\varepsilon_{ij}\,dV\Big)_{\text{hyp elastic}} \approx \Big(\int_{\Omega} \sigma_{ij}\varepsilon_{ij}\,dV\Big)_{\text{elasto-plastic}}, \tag{2}$$

where $U_{\text{hyp elastic}}$ elastic is the total hypothetical elastic strain energy. The pseudo value of the material yield strength $\sigma_{y\ \text{pseudo}}$ is evaluated as follows

$$\frac{U_{\text{dist}}}{U_{\text{hyp elastic}}} \approx \frac{c(\sigma_y - \sigma_{y\ \text{pseudo}})}{\sigma_y}, \tag{3}$$

where U_{dist} is the total strain energy that is vulnerable for redistribution due to plastic deformations and c is a constant that depends on the geometry of the stress-strain curve.

3 Adaptive FEM-BEM coupling method

In this section we present an adaptive FEM-BEM coupling method for elasto-plastic analysis. The method is valid for both 2D and 3D applications. It estimates the FEM and BEM sub-domains and automatically generates/adapts the FEM and BEM meshes/sub-domains, according to the state of computation. The adaptive coupling method improves the estimation of regions where plastic material behaviour is going to develop (regions where the FEM is employed). In the presence of plastic deformations in the FEM region, the solution there is obtained via an iterative scheme. Naturally, an improvement to the estimated FEM and BEM sub-domains will result in additional savings of required system resources and/or a higher potential advantage of eliminating the cumbersome trial and error process in the identification of the FEM and BEM sub-domains. Materials of von-Mises type are considered in this investigation.

The basic steps of implementation of the proposed adaptive FEM-BEM coupling method may be summarized as follows:

1. Levels of loading $(LL_1, LL_2, ..., LL_i, ...LL_m)$ are specified by the user/analyst in order to get an estimate of the FEM and BEM sub-domains (LL_m is the maximum level of loading for the problem at hand).

   If the user/analyst prefers to use a constant interface throughout the FEM-BEM coupling analysis, the maximum load level LL_m is specified (estimated FEM and BEM sub-domains will be utilized for all load increments).
2. For $k = 1,2,...,m$
   2.1. A hypothetical elastic stress state is determined with the load level LL_k via BEM elastic analysis with initial BEM discretization or FEM elastic analysis utilizing a FEM coarse mesh.
   2.2. Regions that violate the yield condition (utilizing the hypothetical elastic stresses of 2.1) are detected. A subsequent elastic analysis is conducted with a "modified" level of loading $LL_{k,\text{mod}}$, "effective"

material properties for the detected regions (effective Young's modulus $E_{k,eff}$ and Poisson ratio $\nu_{k,eff}$) and material properties E and ν for the remainder of the problem.

2.3. The hypothetical elastic state of stresses of step 2.2 is checked against the yield condition. FEM discretization is automatically generated for the regions that violate the yield condition. It may be useful to add a few bands of finite elements around the perimeter of the discretized FEM sub-domain. Consequently, the BEM discretization is generated so as to represent best the remaining bounded/unbounded linear elastic regions (fig. 1).

2.4. Coupled FEM-BEM stress analysis involving elasto-plastic deformations is conducted for the current load increment.

2.5. A repetition of step 2.4 is required for the next load increment if the current state of computation in addition to the load increment is less than or equal to LL_k, else go to step 2.1.

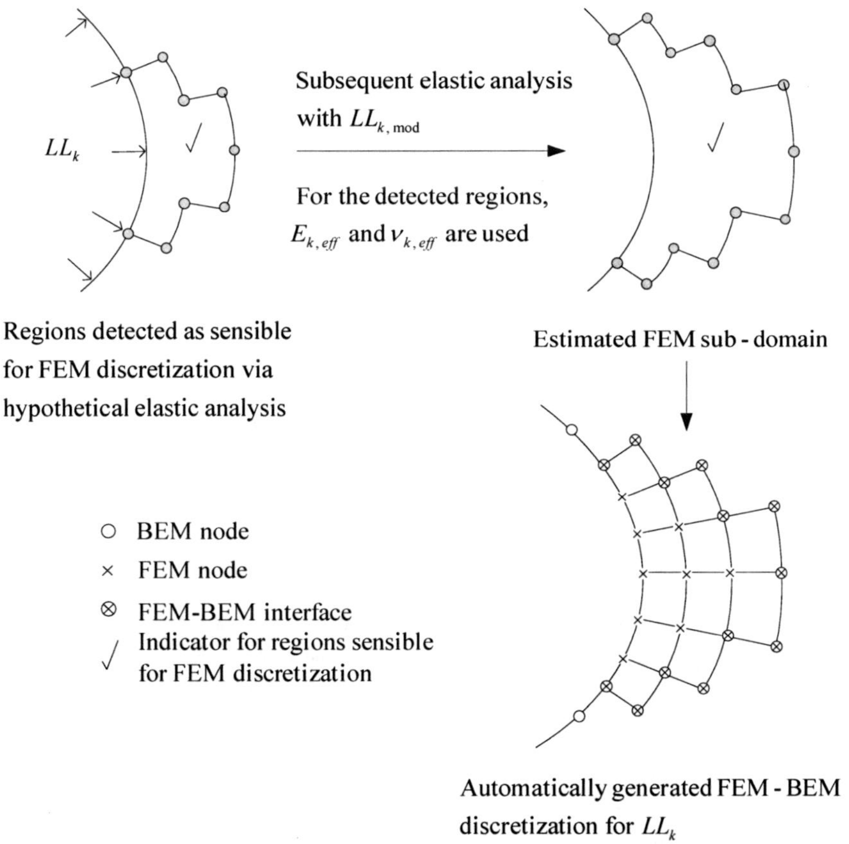

Figure 1: Estimated FEM and BEM sub-domains.

It should be emphasized that steps 1-2.3 are carried out for the sole purpose of estimating and adapting the FEM and BEM sub-domains according to the state of computation. Modified load levels $LL_{k,\,\mathrm{mod}}$ and effective material properties ($E_{k,eff}$ and $\nu_{k,eff}$) are not involved in carrying out step 2.4.

In the remainder of this section we will elaborate more on the determination of the modified level of loading $LL_{k,\,\mathrm{mod}}$ and effective material properties ($E_{k,eff}$ and $\nu_{ki,eff}$) at a typical level of loading LL_k.

The simplicity of linear elastic analysis has motivated some researchers to attempt solving elasto-plastic problems by adapting a modified form of available elastic solutions (see, e.g. references [10,11]). Linear elastic analysis in an iterative manner with a complete spatial distribution of updated material properties is conducted at each iteration in order to approximately simulate elasto-plastic behaviour. The schemes for updating the material properties include projection, arc length, and energy methods (fig. 2). Material points with the same stress level are represented by single points (e.g. a, b, e and f) on the uniaxial stress-strain curve (fig. 3).

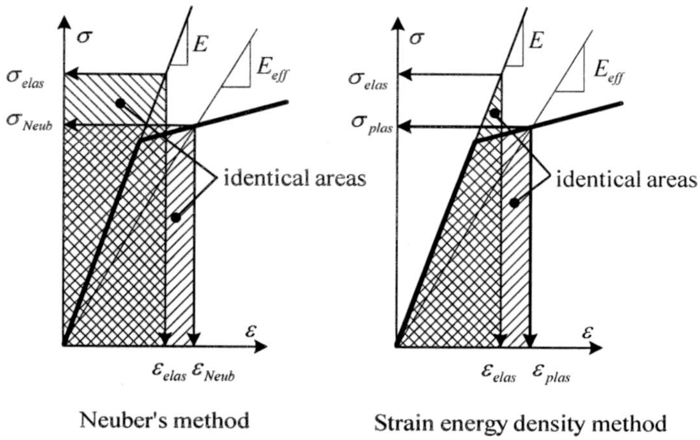

Figure 2: Energetic methods (Neuber and strain energy density methods).

Let us consider materials of von-Mises type obeying a bilinear strain hardening rule. Neuber's method (fig. 2) assumes an energy balance between the strain energy density corresponding to the elasto-plastic stress-strain state and the hypothetical elastic strain energy density (same geometry submitted to the same loading [10,11]). For uni-dimensional states of stress, it is assumed that the product of stress and strain in elasticity is locally identical to the same product calculated by means of an elasto-plastic analysis. For tri-dimensional states of stress, the fundamental hypothesis may be written as [10,11]

$$(\sigma_{ij}\varepsilon_{ij})_{\mathrm{elasto-plastic}} \cong (\sigma_{ij}\varepsilon_{ij})_{\mathrm{hyp\ \ elastic}}. \tag{4}$$

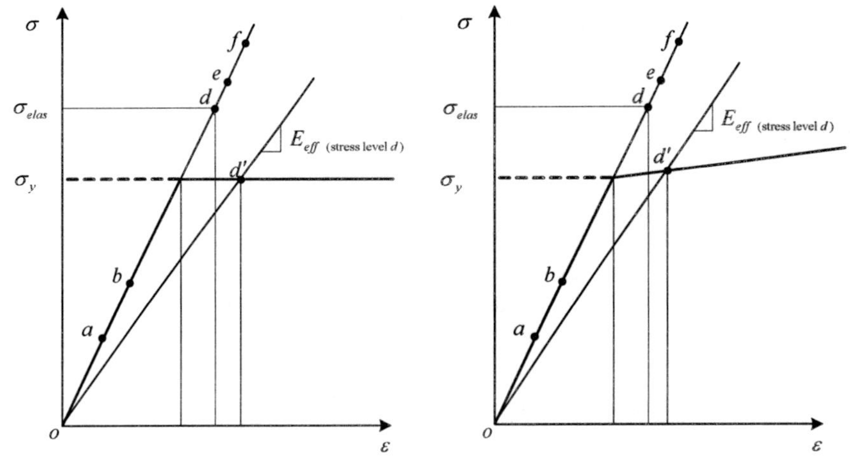

a) Perfect plasticity b) Isotropic hardening

Figure 3: Effective Young's modulus E_{eff}.

From a virtual work principle we utilize the global formulation (eqn (2)). We further assume that there exists a stress level with which effective material properties are determined (stress level d, fig. 3). Next, at a typical level of loading LL_k we define the total strain energy that is vulnerable for redistribution due to plastic deformations $U_{k,\ \mathrm{dist}}$ as the total hypothetical strain energy of the regions that violate the yield condition (step 2.2, basic steps of implementation of the adaptive coupling method)

$$U_{k,\ \mathrm{dist}} = \int_{\Omega} ((\sigma_{ij}\varepsilon_{ij})_{\mathrm{hyp\ \ elastic}})_x \kappa_x dV, \tag{5}$$

where $\kappa_x = 1$ if $((\sigma_{ij}\varepsilon_{ij})_{\mathrm{hyp\ \ elastic}} - \sigma_y\varepsilon_y)_x > 0$, otherwise $\kappa_x = 0$ and σ_y is the uniaxial material yield strength. $\varepsilon_y = \sigma_y / E$. The effective Young's modulus $E_{k,\ eff}$ is then evaluated (fig. 2,3) as follows

$$\frac{U_{k,\mathrm{dist}}}{U_{k,\ \mathrm{hyp\ \ elastic}}} \approx \frac{U_{k,\mathrm{dist\ (stress\ \ level\ d)}}}{U_{k,\mathrm{hyp\ \ elastic\ (stress\ \ level\ d)}}} = \frac{cE_{k,eff}}{E}, \tag{6}$$

where c is a constant. For perfect plasticity and isotropic hardening plasticity models, it is concluded from fig 3 that a reasonable and conservative value is $c = 1$. The effective Poisson ratio $\nu_{k,eff}$ is obtained from equations adopted in iterative elastic analyses in order to simulate elasto-plastic behaviour [10]

$$1/E_{k,eff} = 1/E + 2\phi_k/3 \tag{7}$$

and

$$\nu_{k,eff} = E_{eff}(\nu/E + \phi_k/3). \tag{8}$$

For the determination of the modified level of loading $LL_{k,\text{mod}}$, we further investigate the uniaxial stress-strain curve. Relating the hypothetical elastic stress state curve to that of the effective material parameters, it may be easily concluded that

$$LL_{k,\text{mod}} = LL_i \sqrt{E} \Big/ \sqrt{E_{i,\mathit{eff}}} \,. \tag{9}$$

Compared to the adaptive coupling method of [8,9], the presented method involves additional FEM or BEM elastic iterations. These elastic iterations involved in estimation of the FEM and BEM sub-domains are more than rewarded by an improved estimate of the FEM and BEM sub-domains provided by the proposed adaptive method.

4 Example application

In this section we present an example application that highlights the effectiveness of the adaptive FEM-BEM coupling method presented in Section 3.

The square plate with a centred elliptical defect (fig. 4) is subjected to uniformly distributed tensile loads over the two pairs of the opposing ends (equal biaxaial tension). The applied tractions $P = 100 \text{ x} 10^6$ N/m^2 are scaled with the load factor λ. The elastic material properties of the plate are described by Young's modulus ($E = 206.9\text{x}10^9$ N/m^2) and Poisson's ratio ($\nu = 0.29$). Material of von-Mises type is considered ($\sigma_y = 450\text{x}10^6$ N/m^2), with no hardening effect ($H = 0$), as a yield function and plane strain loading conditions. Due to symmetry, only one quarter of the problem is modelled.

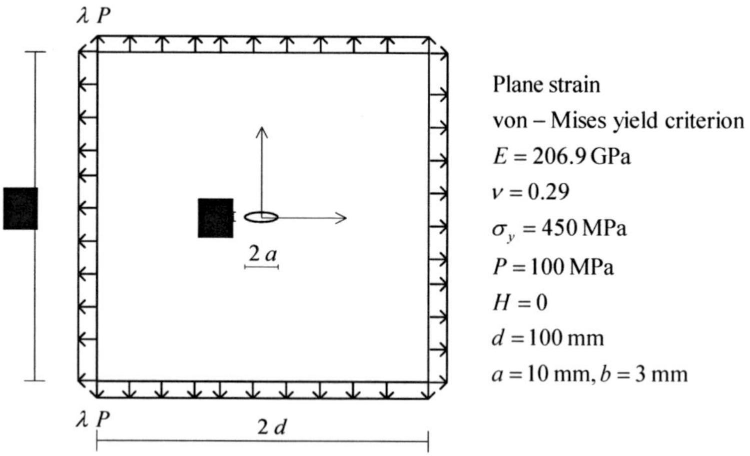

Figure 4: Plate with a centred elliptical defect.

The problem is solved by means of the adaptive coupling method presented in Section 3. The loads are applied incrementally. The elastic prediction and estimates of the regions sensible for discretization by the FEM obtained via the adaptive coupling method are shown in (fig. 5). The coupled FEM-BEM solutions are obtained with the automatically generated FEM and BEM discretization for the particular values of λ. The FEM discretization is generated over the regions that are estimated as sensible for FEM discretization, while the BEM mesh is generated to represent the remaining linear elastic region. Fig. 5 further shows the yielded regions obtained using the adaptive coupled FEM-BEM method for the selected values of λ. The results clearly show that the adaptive FEM-BEM coupled method employs smaller FEM sub-domains. Moreover, the method is practically advantageous as it does not necessitate the predefinition and manual localization of the FEM and BEM sub-domains.

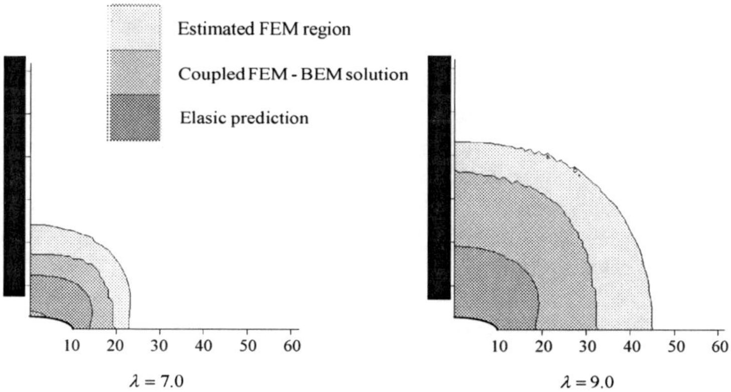

Figure 5: Estimated FEM regions and yielded regions (adaptive FEM-BEM coupling).

5 Conclusion and outlook

The present adaptive coupling method is practically advantageous as it does not necessitate predefinition and manual localization of the FEM and BEM sub-domains. Moreover, the method is computationally efficient as it substantially decreases the size of FEM meshes, which plainly leads to reduction of required system resources and gain in efficiency. The numerical results in 2D elasto-plastic analysis confirm the effectiveness of the proposed method. The extension of the adaptive FEM-BEM coupling method to 3D elasto-plastic applications is currently under investigation. As in the 2D analysis, the FEM sub-domain discretization is progressively adapted and automatically generated to include regions where plasticity occurs, according to the state of computation. Preliminary results indicate the practicality and the efficiency of the FEM-BEM coupling method for 3D elasto-plastic applications.

Acknowledgements

The support of the Austrian Science Fund (FWF), Project number: M950, Lise Meitner Program, is gratefully acknowledged. The authors wish to thank Prof. O.

Steinbach, Graz University of Technology, for providing the symmetric Galerkin boundary element computer code utilized in some parts of this investigation.

References

[1] Astrinidis, E., Fenner, R. T. & Tsamasphyros, G., Elastoplastic Analysis with Adaptive Boundary Element Method. *Computational Mechanics*, **33(3)**, pp. 186–193, 2004.

[2] Maischak, M. & Stephan, E. P., Adaptive hp-Versions of Boundary Element Methods for Elastic Contact Problems. *Computational Mechanics*, **39(5)**, pp. 597–607, 2007.

[3] Ribeiro, T. S. A., Beer, G. & Duenser, C., Efficient elastoplastic analysis with the boundary element method. *Computational Mechanics*, **41(5)**, pp. 715–732, 2008.

[4] Brink, U., Klaas, O., Niekamp, R. & Stein, E., Coupling of Adaptively Refined Dual Mixed Finite Elements and Boundary Elements in Linear Elasticity. *Advances in Engineering Software*, **24(1–3)**, pp. 13–26, 1995.

[5] Carstensen, C., Zarrabi, D. & Stephan, E. P., On the h-Adaptive Coupling of FE and BE for Viscoplastic and Elasto-Plastic Interface Problems. *Journal of Computational and Applied Mathematics*, **75(2)**, pp. 345–363, 1996.

[6] Mund, P. & Stephan, P., An Adaptive Two-Level Method for the Coupling of Nonlinear FEM-BEM Equations. *SIAM Journal on Numerical Analysis*, **36(4)**, pp. 1001–1021, 1999.

[7] Doherty, J. P. & Deeks, A. J., Adaptive Coupling of the Finite-Element and Scaled Boundary Finite-Element Methods for Non-Linear Analysis of Unbounded Media. *Computers and Geotechnics*, **32(6)**, pp. 436–444, 2005.

[8] Elleithy, W. & Langer, U., "Adaptive FEM-BEM Coupling Method for Elasto-Plastic Analysis," Proc. of the 8th Int. Conf. on Boundary Element Techniques, eds. V. Minutolo & M.H. Aliabadi, EC: UK, pp. 269–274, 2007.

[9] Elleithy, W. M., Analysis of Problems in Elasto-Plasticity via an Adaptive FEM-BEM Coupling Method. *Computer Methods in Applied Mechanics and Engineering*, in press, doi:10.1016/j.cma.2008.02.018, 2008.

[10] Jahed, H., Sethuraman, R. and Dubey, R. N., A Variable Material Property Approach for Solving Elasto-Plastic Problems. *International Journal of Pressure Vessels and Piping*, **71(3)**, pp. 285–291, 1997.

[11] Desmorat, R., Fast Estimation of Localized Plasticity and Damage by Energetic Methods. *International Journal of Solids and Structures*, **39(12)**, pp. 3289–3310, 2002.

3D multi-spectral image-guided Near-Infrared spectroscopy using the boundary element method

S. Srinivasan, B. W. Pogue & K. D. Paulsen
Thayer School of Engineering, Dartmouth College, Hanover, NH-03755, USA

Abstract

Image guided (IG) Near-Infrared spectroscopy (NIRS) has the ability to provide high-resolution metabolic and vascular characterization of tissue, with clinical applications in the diagnosis of breast cancer. This method is specific to multi-modality imaging where tissue boundaries obtained from alternate modalities such as MRI/CT, are used for NIRS recovery. IG-NIRS is severely limited in 3D by challenges such as volumetric meshing of arbitrary anatomical shapes and the computational burden encountered by existing models that use the finite element method (FEM). We present an efficient and feasible alternative to the FEM using the boundary element method (BEM). The main advantage is the use of surface discretization, which is reliable and more easily generated than volume grids in 3D and enables automation for large numbers of clinical data-sets. The BEM has been implemented for the diffusion equation to model light propagation in tissue. Image reconstruction based on the BEM has been tested in a multi-threading environment using four processors, which provide 60% improvement in computational time compared to a single processor. Spectral priors have been implemented in this framework and applied to a three-region problem with a mean error of 6% in recovery of NIRS parameters.

Keywords: NIR spectroscopy, image guided, boundary element method.

1 Introduction

NIR optical imaging has the ability to provide non-invasive functional characterization of tissue relating to its metabolic and vascular status [1]. This yields unique intrinsic information regarding the composition of tissue,

WIT Transactions on Modelling and Simulation, Vol 47, © 2008 WIT Press
www.witpress.com, ISSN 1743-355X (on-line)
doi:10.2495/BE080241

not available from conventional imaging modalities such as MRI, x-rays or ultrasound, which depend on structural features for contrast. NIR spectroscopy and tomography have been used to study breast tissue for the diagnosis of breast cancer [2], monitoring neoadjuvant chemotherapy [3] and assessing risk of cancer through screening [4]; as well as in other areas such as the study of brain physiology [5], muscle [6] etc. Our particular interest is in exploring the use of NIR imaging to improve the specificity of breast cancer diagnosis [7].

The high scattering in tissues like the breast allow penetration of light up to a dozen centimeters, crucial to deep-tissue imaging; however this same scattering also limits the spatial resolution of NIR imaging to millimeter scale. Over the past decade, research has explored the synergy of hybrid multi-modality imaging by combining modalities such as MRI, ultrasound or x-ray with NIR to increase the spatial resolution and improve sensitivity and specificity of breast cancer diagnosis [8–10]. At Dartmouth, a hybrid MRI-NIR system has been developed for this purpose [8].

The image reconstruction algorithms developed for hybrid systems have studied various intelligent ways of incorporating the anatomical information from MRI, x-ray or CT, into the ill-posed NIR image recovery process [11, 12]. The most intuitive and accurate results have been obtained using the *'hard priors'* algorithm which assumes the domain to contain piece-wise constant regions whose exact boundaries are obtained from the alternate modality such as MRI [11, 13]. This produces the best global estimates of NIR vascular and structural parameters such as total hemoglobin, oxygen saturation, water, scatterer size and density. Currently, FEM techniques are utilized for this purposes which have severe drawbacks in 3D IG-NIR reconstructions namely: (1) need for volume discretization of arbitrary breast shapes which is a complex procedure and does not lend easily to automation of large clinical data sets, and (2) requirement of large meshes for accuracy (>30 000 nodes) of the FEM method, though reconstructing for only few parameters (~10-15 values): this has implications in computational time and efficiency. To overcome these issues, we have developed BEM for IG-NIRS reconstructions.

BEM is an attractive alternative numerical method requiring only surface discretization, which are faster and more reliably produced as compared to volume meshes. BEM uses the assumption that the imaging domain contains known number of piece-wise constant regions whose boundaries are obtained from MRI, which is the same as *hard priors*. The BEM is uniquely suited for IG-NIRS, since the breast typically consists of limited tissue types such as adipose, fibroglandular tissues and the lesion (malignant and/or benign), and the exact knowledge of these tissue layers can be obtained from MRI. Towards this end, we have developed a forward model to the diffusion equation based on the BEM [14], which solves for fluence given a distribution of optical properties in the tissue domain. Image reconstruction involves solving the inverse problem of estimating optical properties of the tissue types given a set of measurements of fluence in the domain; this has been implemented using the Newton's method. Addition of spectral constraints to reconstruct for wavelength-independent parameters have significant impact on the accuracy of total hemoglobin, oxygen saturation, water and scattering [7, 15]; and have been implemented in the BEM framework. We present application of this 3D BEM-based multi-spectral IG-NIRS in simulations and discuss future applications in clinical setting.

2 Methods

The forward model is represented by the diffusion equation involves obtaining the light flux outgoing from different surfaces, using known optical properties for the interior of these regions. The diffusion approximation to the radiative transport equation assumes that the interior photon irradiance is highly scattered and nearly uniform in all directions, and therefore, its angular distribution is effectively described by the single isotropic fluence parameter, Φ [16], and is given by:

$$-\nabla.D(r)\nabla\Phi(r,\omega)+\left(\mu_a(r)+\frac{i\omega}{c}\right)\Phi(r,\omega)=q_0(r,\omega) \tag{1}$$

When tissue is assumed to consist of piece-wise constant homogeneous regions, the diffusion coefficient, *D(r)* is constant in each zone so that the diffusion equation can be expressed in the form of a modified Helmholtz equation,

$$\nabla.D_l\nabla\Phi-k_l^{\ 2}\Phi=-q_0(r,\omega) \tag{2}$$

where k_l is the wave number, constant in sub-domain l [17, 18], and for a particular frequency ω, $k_l^{\ 2}=\left(\mu_a(r)+\frac{i\omega}{c}\right)$.The fundamental Green's function solution is readily available for this equation and can be used to derive the BEM formulation[19] given as:

$$c_i\Phi_i+\oint D_l\frac{\partial G_i}{\partial n}\Phi-\oint D_l\frac{\partial\Phi}{\partial n}G_i=\langle q_0,G_i\rangle \tag{3}$$

where Gi is the Green's function for the modified Helmholtz equation given in 3D as:

$$G_i(r,r_i)=\frac{\exp\left(\frac{-k_l|r-r_i|}{\sqrt{D_l}}\right)}{4\pi D_l|r-r_i|}\ ;\quad c_i=\left\{\frac{\Omega}{4\pi}\right.$$

and Ω is the solid angle enclosed by the boundary at node i [19] and $\oint$ is the integral over the boundary. Discretizing equation 2 through the linear basis function ψ and reformulating the equation in matrix form results in:

$$[A]\{\Phi_i\}-[B]\left\{D_l\frac{\partial\Phi_i}{\partial n}\right\}=\{Q_i\} \tag{4}$$

where: $A_{i,j}=c_i\delta_{ij}+\oint D_l\frac{\partial G_i}{\partial n}\psi_j ds$, $B_{ij}=\oint G_i\psi_j ds$ and $Q_i=\langle q_0,G_i\rangle$.

A point source representation was implemented for our purpose. A type III boundary condition on the outer boundary was used given by:

$$\Phi_{al} + \frac{D_l}{\alpha} \frac{\partial \Phi}{\partial n}\bigg|_{al} = 0 \tag{5}$$

The method can be suitably extended to multiple regions by implementing continuity conditions across inner boundaries. Further details can be found elsewhere [14]. The forward model has been implemented in matlab using mex files in fortran for building the matrices and tested in simulations. Results [14] from application in two-region problems indicated that BEM was more accurate than FEM at a lower resolution; and showed 44–72% improvement in computational time over FEM. The forward model was also demonstrated on a patient-specific mesh generated from the MRI of a mammographically normal subject [20].

Image reconstruction of optical properties was implemented based on this forward model to the diffusion equation. A modified iterative Newton's method was used for the reconstruction procedure where the least squares norm of the difference between measured data (from the NIR system) and the numerical model data was minimized [21]. A perturbation approach was used to calculate the sensitivity of the boundary measurements to change in optical properties. Regularization was used to counter the ill-conditioned nature of the problem [22] and a stopping criterion of less than 0.5% change in the error between successive iterations was applied. All reconstructions in this work were run in a multi-threading environment on a Beowulf cluster using four processors.

3 Results

BEM-based recovery of optical properties was demonstrated previously in an experimental setting and using patient-specific mesh [14]. Recently, we have implemented the BEM to run in a multi-threaded environment using Matlab Distributive computing toolbox. In order to compare the speed-up in computational time, simulated data was generated on the imaging domain shown in Figure 1(a). The domain was described using a surface grid containing 4126 nodes generated using Netgen [23]. The measurement geometry comprised of 16 sources with 15 detectors per source (=240 measurements) placed in a ring around the central plane of the domain. 1% random Gaussian noise was added to these measurements and used to reconstruct for optical properties using a coarser surface mesh containing 2780 nodes. The recovered values for absorption and reduced scattering coefficients are shown in Figures 1(b) and (c). The recovered absorption in the inclusion was accurate within 89% and scattering was accurate within 99%. The time of computation for the reconstruction using a single processor versus four processors is shown in Figure 1(d). The use of four processors provided an improvement of 59% in computational time. In all the following work, we have utilized four processors for BEM parallel computing.

Spectral priors: The known spectral signatures of the primary absorbers in tissue (see Figure 2(a)) can be embedded into the reconstruction to use multi-

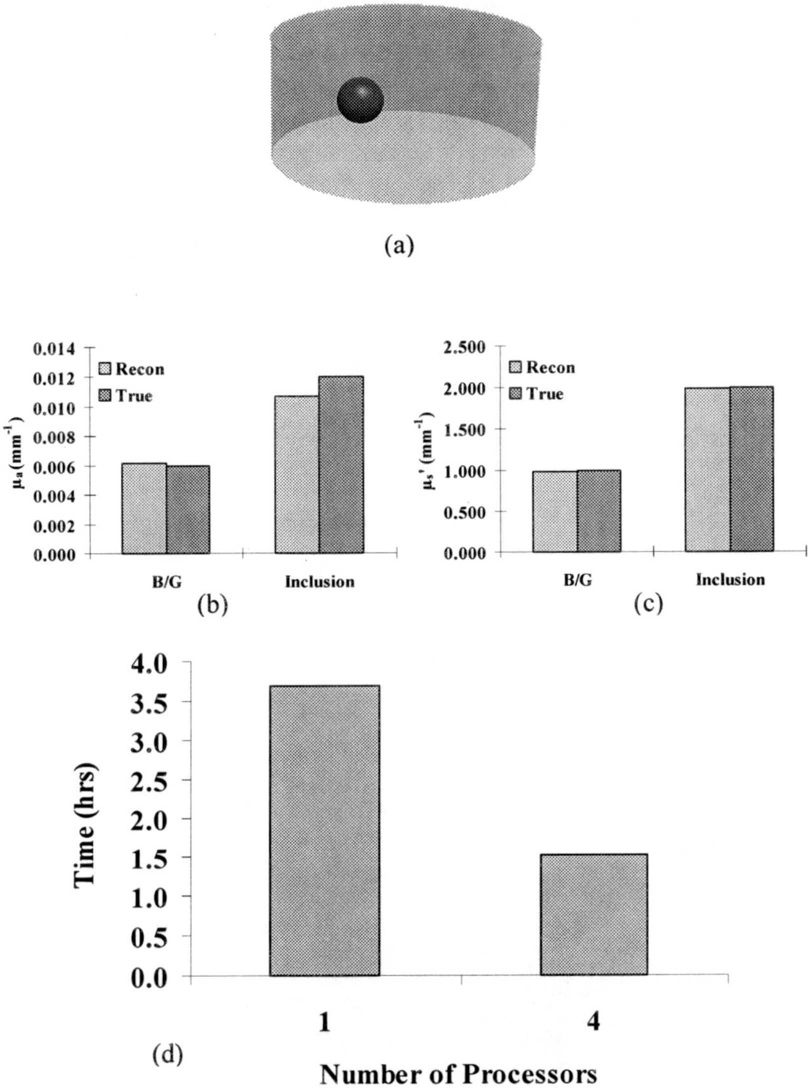

Figure 1: (a) Test imaging domain used to create a surface mesh for generating forward data. The data with 1% noise was reconstructed using a coarser surface grid and recovered values are shown for absorption coefficient (b) and reduced scattering coefficient (c). (d) The differences in computational time using a single versus four processors in a multi-threaded environment are shown.

wavelength measurements for direct recovery of chromophore concentrations (such as oxy and deoxyhemoglobin and water) and scattering parameters. This has been shown to improve the recovery of these functional parameters in the FEM framework [15] and we have implemented them in the BEM framework here. To test 3-D spectral BEM reconstruction, measurements at six NIR wavelengths were generated for a three-region test domain (shown in Figure 2(b)) with appropriate concentrations[7] to mimic the adipose, fibroglandular and tumor regions. 1% random Gaussian noise was added to these measurements and used to reconstruct for the chromophore concentrations and scattering. The recovered values are shown in Figure 3(a) and (b). The computational time was 2.15 hrs using 4 processors. The mean error in the reconstructed values over all parameters was calculated to be 0.2% in region 1, 4% in region 2 and 13% in region 3. This shows the feasibility of using BEM for IG-NIRS of multi-layered domains in 3-D.

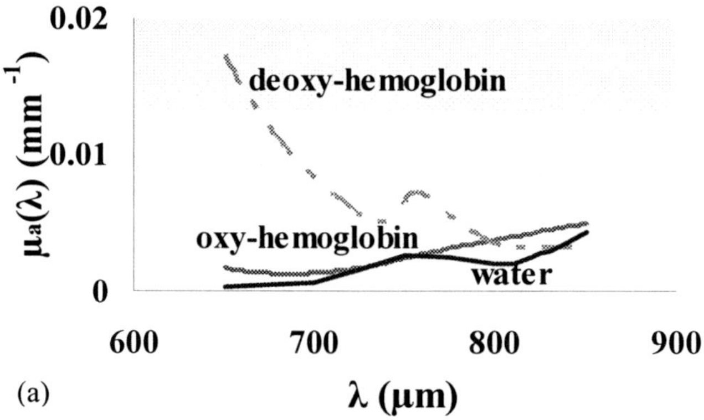

(a)

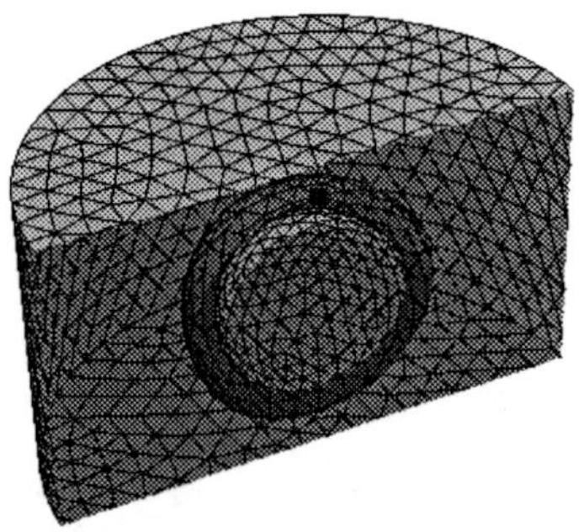

(b)

Figure 2: (a) Known spectral signatures of oxyhemoglobin, de-oxyhemoglobin and water obtained from literature, can be incorporated into image reconstruction. (b) Cross-section of surface grid generated for a three-region domain, to model the multiple layers of adipose, fibroglandular and tumor layers in a breast. Appropriate concentrations were used to generate forward data.

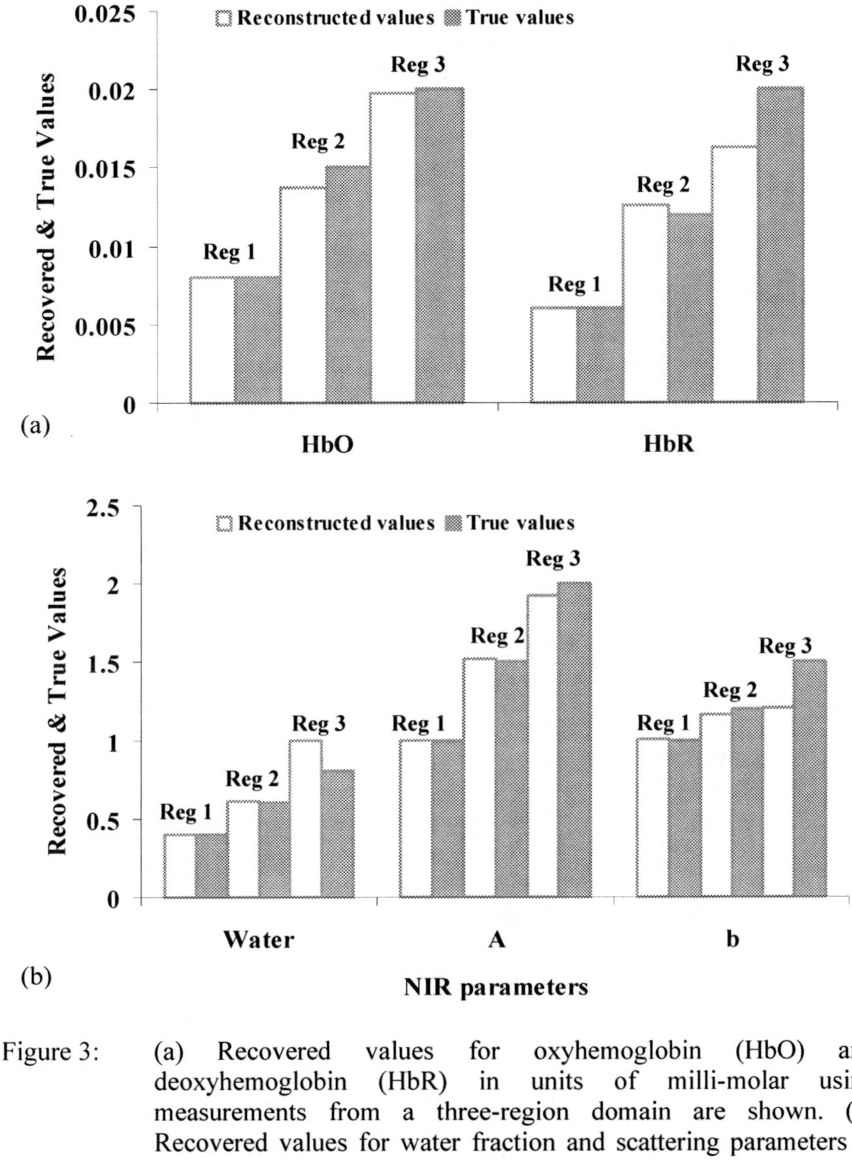

Figure 3: (a) Recovered values for oxyhemoglobin (HbO) and deoxyhemoglobin (HbR) in units of milli-molar using measurements from a three-region domain are shown. (b) Recovered values for water fraction and scattering parameters A and b derived from the empirical relationship $\mu_s^{'}(\lambda) = A\lambda^{-b}$.

4 Conclusions

A BEM approach for 3-D multi-spectral IG-NIRS has been developed and provides an efficient technique for computing in 3-D using only surface discretization. Image reconstruction using BEM was run in a multi-threaded environment using 4 processors and showed ~60% improvement in computational time as compared to use of a single processor. We are currently limited by the number of processors and will extend it to 8 processors, planned in the future. Spectral priors were incorporated in this framework and used to reconstruct directly for chromophore concentrations and scattering in a multi-layered media; results were accurate with overall mean error of 6%. This method will be used to reconstruct for 3-D tissue IG-NIR spectroscopic measures using clinical in-vivo measurements at multi-wavelengths obtained from hybrid MR-NIR system based at Dartmouth.

References

[1] Jobsis, F.F., Non-invasive, infra-red monitoring of cerebral and myocardial oxygen sufficiency and circulatory parameters. *Science,* 1977. **198**: p. 1264–1267.

[2] Pogue, B.W., S.P. Poplack, T.O. McBride, W.A. Wells, K.S. Osterman, U.L. Osterberg, and K.D. Paulsen, Quantitative Hemoglobin Tomography with Diffuse Near-Infrared Spectroscopy: Pilot Results in the Breast. *Radiology,* 2001. **218**(1): p. 261–6.

[3] Jakubowski, D.B., A.E. Cerussi, F. Bevilacqua, N. Shah, D. Hsiang, J. Butler, and B.J. Tromberg, Monitoring neoadjuvant chemotherapy in breast cancer using quantitative diffuse optical spectroscopy: a case study. *J Biomed Opt,* 2004. **9**(1): p. 230–8.

[4] Simick, M.K., R. Jong, B. Wilson, and L. Lilge, Non-ionizing near-infrared radiation transillumination spectroscopy for breast tissue density and assessment of breast cancer risk. *Journal of Biomed Opt,* 2004. **9**(4): p. 794–803.

[5] Strangman, G., D.A. Boas, and J.P. Sutton, Non-invasive neuroimaging using near-infrared light. *Biol. Psychiatry,* 2002. **52**: p. 679–693.

[6] Gratton, E., S. Fantini, M.A. Franceschini, G. Gratton, and M. Fabiani, Measurements of scattering and absorption changes in muscle and brain. *Phil. Trans. R. Soc. Lond. B,* 1997. **352**: p. 727–735.

[7] Srinivasan, S., B.W. Pogue, B. Brooksby, S. Jiang, H. Dehghani, C. Kogel, W.A. Wells, S.P. Poplack, and K.D. Paulsen, Near-infrared characterization of breast tumors in-vivo using spectrally-constrained reconstruction. *Technology in Cancer Research and Treatment,* 2005. **4**(5): p. 513–526.

[8] Carpenter, C., B.W. Pogue, S. Jiang, H. Dehghani, X. Wang, K.D. Paulsen, W.A. Wells, J. Forero, C. Kogel, J. Weaver, S.P. Poplack, and P.A. Kaufman, Image-guided optical spectroscopy provides molecular-specific information in vivo: MRI-guided spectroscopy of breast cancer hemoglobin, water & scatterer Size. *Optics Letters,* 2007. **32**(8): p. 933–935.

[9] Zhang, Q., T.J. Brukilacchio, A. Li, J.J. Stott, T. Chaves, E. Hillman, T. Wu, M. Chorlton, E. Rafferty, R.H. Moore, D.B. Kopans, and D.A. Boas, Coregistered tomographic x-ray and optical breast imaging: initial results. *J Biomed Opt,* 2005. **10**(2): p. 024033–0240339.

[10] Zhu, Q., E. Conant, and B. Chance, Optical imaging as an adjunct to sonograph in differentiating benign from malignant breast lesions. *Journal of Biomedical Optics,* 2000. **5**(2): p. 229–36.

[11] Brooksby, B., Dehghani, H., Pogue, B. W., Paulsen, K. D., Near infrared (NIR) tomography breast image reconstruction with apriori structural information from MRI: algorithm development for reconstructing heterogeneities. *IEEE Journal of Selected Topics in Quantum Electronics on Lasers in Medicine and Biology,* 2003. **9**: p. 199–209.

[12] Boverman, G., E.L. Miller, A. Li, Q. Zhang, T. Chaves, D.H. Brooks, and D. Boas, Quantitative spectroscopic diffuse optical tomography of the breast guided by imperfect a priori structural information. *Phys Med Biol,* 2005. **50**: p. 3941–3956.

[13] Dehghani, H., B.W. Pogue, J. Shudong, B. Brooksby, and K.D. Paulsen, Three-dimensional optical-tomography: resolution in small-object imaging. *Applied Optics,* 2003. **42**(16): p. 3117–3128.

[14] Srinivasan, S., B.W. Pogue, C. Carpenter, P.K. Yalavarthy, and K.D. Paulsen, A boundary element approach for image-guided near-infrared absorption and scatter estimation. *Medical Physics,* 2007. **34**(11): p. 4545–57.

[15] Srinivasan, S., Pogue, B. W., Jiang, S., Dehghani, H. and Paulsen, K. D., Spectrally constrained chromophore and scattering NIR tomography provides quantitative and robust reconstruction. *Applied Optics,* 2005. **44**(10): p. 1858–69.

[16] Ishimaru, A., *Wave propagation and scattering in random media.* Vol. 1. 1978: Academic Press, Inc., New York.

[17] Fedele, F., M.J. Eppstein, J.P. Laible, A. Godavarty, and E.M. Sevick-Muraca, Fluorescence photon migration by the boundary element method. *Journal of computational physics,* 2005. **210**: p. 109–132.

[18] Sikora, J., A. Zacharopoulos, A. Douiri, M. Schweiger, L. Horesh, S.R. Arridge, and J. Ripoll, Diffuse photon propagation in multilayered geometries. *Phys Med Biol,* 2006.

[19] Brebbia, C.A. and J. Dominguez, *Boundary elements: an introductory course.* II Edition, 1992: Computational Mechanics Publications & McGraw-Hill Book Company.

[20] Srinivasan, S., C. Carpenter, B.W. Pogue, and K.D. Paulsen, 3D Image-guided absorption and scatter tomography using boundary element method. *Biomedical Optics, OSA Technical Digest,* 2008: p. BSuE36.

[21] Paulsen, K.D., and Jiang, H., Spatially varying optical property reconstruction using a finite element diffusion equation approximation. *Med. Phys.,* 1995. **22**(6): p. 691–701.

[22] Marquardt, D.W., An algorithm for least squares estimation of nonlinear parameters. *J. Soc. Ind. Appl. Math.,* 1963. **11**: p. 431–441.

[23] Schöberl, J., NETGEN - An advancing front 2D/3D-mesh generator based on abstract rules.

Parametric analysis of ACA-based solver for BEM 3D

T. Grytsenko & A. Peratta
Wessex Institute of Technology, UK

Abstract

The standard Boundary Element Method (BEM) with single domain yields the dense linear system of equation (LSE). This LSE cannot be solved efficiently by existing iterative solvers such as CG, GMRES, TFQMR, etc. Moreover, such LSE cannot be stored in advanced formats such as Compressed Sparse Row or Modified Sparse Row and therefore cannot be compressed in order to economise memory consumption. Methods such as the fast multipole method and panel clustering gain their efficiency from approximating the kernel function. This paper implements and studies the parameters and performance of the Adaptive Cross Approximation (ACA) algorithm for reducing the rank of off-diagonal blocks in the three dimensional BEM. The ACA algorithm is purely algebraic and does not need to deal with the kernel. The algorithm uses a hierarchical matrix storage approach where the matrix is split into many blocks. The off-diagonal blocks of this matrix represent remote interactions between source points and are therefore approximated by low-rank matrices with the ACA approach. Those blocks, which describe close interactions between source points, are stored without any changes. These reorganisations in conjunction with novel algorithms for manipulation with H-matrices reduce the calculation complexity of matrix-vector multiplication (MVM) to approximately $O(N)$. This simplification of MVM paves the road for speeding up of the solution of LSE coming from BEM, as MVM is a key-operation for many existing solvers. This paper analyses the ACA-based solver and describes how to choose parameters for this solver when used in conjunction with BEM. Finally, a 3D numerical example solved with this technique is presented.
Keywords: BEM 3D, hierarchical matrix, adaptive cross approximation, iterative solver.

WIT Transactions on Modelling and Simulation, Vol 47, © 2008 WIT Press
www.witpress.com, ISSN 1743-355X (on-line)
doi:10.2495/BE080251

1 Introduction

The classical BEM method yields a dense non-symmetric linear system of equations (LSE). This LSE cannot be solved efficiently by existing iterative solvers such as CG, GMRES, TFQMR, etc, which exploit the sparse pattern of the matrix. Moreover, it is inefficient to store such dense LSE in sparse formats such as Compressed Sparse Row (CSR) or Modified Sparse Row (MSR). In contrast, other methods such as the fast multipole method and panel clustering gain their efficiency from approximating the kernel function [1,2], yielding a sparse LSE. This paper implements and studies the parameters and performance of the Adaptive Cross Approximation (ACA) algorithm [5] for reducing the rank of off-diagonal blocks in the three dimensional BEM matrix. The great advantage of the ACA algorithm is that it is purely algebraic and does not need to deal with the kernel. The algorithm uses a hierarchical matrix storage approach where the matrix is split into many blocks classified into two categories, weakly and strongly coupled. The former are off-diagonal blocks, which represent remote interactions between source points and field elements, and therefore can be approximated by low-rank matrices. For this purpose Adaptive Cross Approximation (ACA) algorithm is used [5]. These blocks are stored in a special Rk-format. The latter blocks describe close interactions between source points, and field elements are stored without any changes in a full-matrix format. This reorganisation of the LSE, implemented in conjunction with novel algorithms for manipulation with H-matrices, reduces the calculation complexity of matrix-vector multiplication (MVM) to approximately $O(N)$ [2]. The simplification of MVM considerably reduces the computational burden of the solving stage.

This paper is organised as follows: in Section 2, key-points of H-matrices technique are given, particularly their structure, cluster tree, domain partitioning and admissibility conditions; in Section 3, two versions of ACA-algorithm are described; Section 4 analyses ACA-algorithm and describes how to choose parameters for this solver; Section 5 represents numerical results; Section 6 makes the conclusions; our acknowledgements are expressed in the final section.

2 Hierarchical matrices

This section provides a brief introduction to hierarchical matrices and explains their structure and creation. The concepts of mesh partitioning, cluster tree structure and admissibility condition [2–4] are also described.

2.1 Mesh partitioning and its storage scheme

The main task of mesh partitioning scheme is to build a hierarchy of source points (indices) according to geometrical criterion. The indices I of nodes are stored in the so-called cluster tree [2] (see Figure 1). The root of the tree T_I is the index set $T_I = \{0,..., N-1\}$ where N is the number of degrees of freedom.

The subsequent node t with more than N_{ind} indices, where N_{ind} is a parameter, has exactly two successors: the first contains the first half of its indices and the second one the second half. Nodes with no more than N_{ind} indices are leaves. There are many ways to build such a partition [6]. In this work geometric a bisection algorithm has been used. Cluster tree for a circle with 10 degrees of freedom (DOF) and $N_{ind} = 2$ is represented in Figure 1.

In Figure 1 notation $I_{5(2)}^{(3)}$ means that node number 5 is situated on third level of cluster tree and consists of two indices or degrees of freedom. Cluster tree is a basis for construction of H-matrix.

2.2 Structure of hierarchical matrices

In general case H-matrix represents hierarchy of Rk-matrices and Full-matrices [2,4]. An $n \times m$ matrix M of rank at most k is said to be stored in Rk-matrix representation if it is stored in factorised form $M = AB^T$ where matrices $A \in R^{n \times k}$ and $B \in R^{m \times k}$ are both stored as an array (column-wise). In C programming language it looks as in Figure 2 (i).

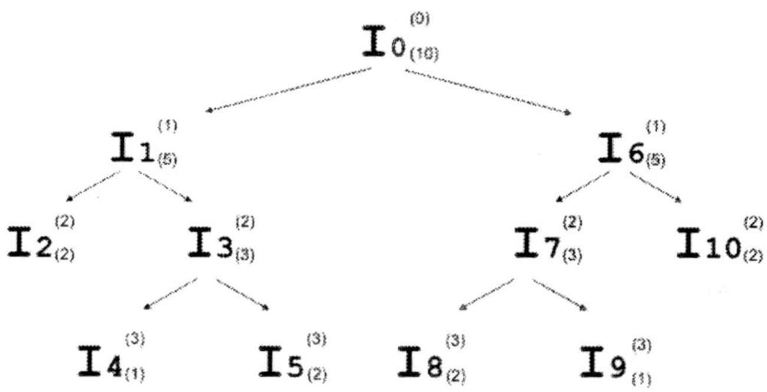

Figure 1: Cluster tree ($N=10$, $N_{ind} = 2$).

(i) – Rk- matrix	(ii) – Full-matrix	(iii) – Super matrix

`struct rkmatrix {`
`Int k;`
`Int rows;`
`Int cols;`
`Double * a;`
`Double * b;`
`};`
	`struct fullmatrix`
	`{`
	`Int rows;`
	`Int cols;`
	`Double * e;`
	`};`
		`struct`
		`supermatrix {`
		`Int rows;`
		`Int cols;`
		`Int block_rows;`
		`Int block_cols;`
		`Prkmatrix r;`
		`Pfullmatrix f;`
		`Psupermatrix * s;`
		`};`

Figure 2: Basic structures in H-matrix theory.

An $n \times m$ matrix M is said to be stored in Full matrix representation if the entries M_{ij} are stored as real numbers in an array of length $n*m$ in the order $M_{11},...,M_{n1},M_{12},...,M_{n2},...,M_{1m},...,M_{nm}$. In C programming language it looks as in Figure 2 (ii).

The matrix M is said to be stored in H-matrix representation if the submatrices or blocks are stored either in Full-matrix or Rk-matrix representation. Each block in the H-matrix tree can be: (i) a leaf: then the corresponding matrix block is represented by Full matrix or Rk-matrix; (ii) a node: then the block is decomposed into sons. This means that the matrix corresponding to this block is a supermatrix that consists of submatrices corresponding to its sons. In C programming language it looks as in Figure 2 (iii).

2.3 Admissibility condition

Each block in H-matrix represents interactions of nodes in a cluster tree, particularly interactions of their DOFs. Those nodes of cluster tree which are geometrically situated far one from the other form the blocks in Rk-format and those that situated closely yields the blocks in Full-format. There is a special criterion in order to find how far one from the others nodes are:

$$dist(C_{N1}, C_{N2}) \geq \gamma * Max\_diam(C_{N1}, C_{N2}), \tag{1}$$

where C_{N1}, C_{N2} are clusters in a cluster tree; γ is a coefficient and $Max\_diam(C_{N1}, C_{N2})$ is a function that finds maximal diameter among clusters C_{N1}, C_{N2}. This criterion calls admissibility condition [4]. Those clusters that are admissible according to (1) form blocks in Rk-format and those which are not generate blocks in Full-format.

Adaptive cross approximation (ACA)-algorithm described in the next section is developed for low-rank approximation of large matrices. This algorithm works with whole H-matrix. If a block is not admissible then its entries are stored without approximation, otherwise ACA can be applied. So, ACA approximates only admissible block, i.e. blocks stored in Rk-format.

3 Adaptive cross approximation

The off-diagonal blocks of H-matrix describe remote interactions between source points and can be approximated by low rank matrices [5]. ACA algorithm is developed to generate low rank approximants for such blocks. In contrast to other methods such as fast multipole, panel clustering, etc., the low-rank approximant is not generated by replacing the kernel function of the integral operator. This algorithm is fully algebraic and uses the original matrix entries to compute the low-rank approximant. ACA algorithm does not change kernel function and can be applied only to the discrete integral operators with asymptotically smooth kernels.

There are two versions of ACA algorithm: fully and partially pivoted. Fully pivoted realisation is feasible in a case when coefficient matrix $\mathbf{A}$ of the equation

$$\mathbf{A}\,\mathbf{x} = \mathbf{b} \tag{2}$$

is already computed. In this case, ACA in conjunction with H-matrix technique are used to speedup solving of LSE. However, if matrix A is not yet computed and there is a possibility to generate its entries individually there is no need to compute whole matrix beforehand. Instead, ACA approximation can be integrated into computation stage in order to approximate this matrix on the fly. In this case not all entries of whole matrix must be calculated. For this purpose partially pivoted ACA algorithm can be used. Detailed description can be found in [5].

There is an error of ACA-approximation ε_{ACA}, which defines how deep the approximation is. $\varepsilon_{ACA} = 0$ means that there is no approximation, any $\varepsilon_{ACA} > 0$ leads to changes in the original matrix A (2). By using ACA algorithm any $\varepsilon_{ACA} \geq 0$ may be reached. The memory capacity needed for storing an approximant depends on ε_{ACA}: higher error will need less memory. However, high ε_{ACA} may damage LSE and as a result will lead to a very coarse result, especially if the system is ill-conditioned. So, selection of ε_{ACA} is a matter of balance between memory capacity, CPU time and precision of the result x of the equation (2).

4 Selection of ACA parameters

It is desirable that the auxiliary parameters controlling the behaviour of a solver are adjusted in order to optimise: CPU time T_{CPU}, memory consumption M and the solution error ε_{sol}. ACA-based solver involves some auxiliary parameters that play a key role in the optimisation of those main characteristics: γ from the equation (1), size of cluster N_{ind} in cluster tree, ACA-error ε_{ACA} and number of DOFs N.

In order to find optimal values for each parameter it is necessary to define dependencies between them. The most important parameter is the solution error, which depends on three auxiliary parameters $\gamma, N_{ind}, \varepsilon_{ACA}$. Different values of γ and N_{ind} yield different cluster tree and as a result different H-matrix:

$$K_{BL}(N, N_{ind}) = a + b\,N + c\,N^2 + d\,/\,N_{ind} \tag{3}$$

where $K_{BL} = N_{RK} / N_{FULL}$ is the ratio between the number of Rk-blocks and Full-blocks in H-matrix, and *a*, *b*, *c*, *d* are fitting coefficients. This relationship and others in this paper have been established by means of non-linear least-squares method for the model described in Section 5. Depending on parameter γ the fitting coefficients change as follows:

$$K_{BL} = 0.0257 + 1e-04\,N + 8.65e-09\,N^2 + \\ +33.37 / N_{ind} \quad for\ \gamma = 0.5 \tag{4}$$

$$K_{BL} = 0.725 + 9.83e-05 * N + 1.34e-09 * N^2 + \\ +36.98 / N_{ind} \quad for\ \gamma = 1.0 \tag{5}$$

$$K_{BL} = 0.492 + 2.2e-04 * N - 3.13e-08 * N^2 + \\ +68.98 / N_{ind} \quad for\ \gamma = 2.5 \tag{6}$$

The higher γ yields higher K_{BL} and as a result higher solution error:

$$\ln \varepsilon_{sol}(K_{BL}, \varepsilon_{ACA}) = -4.21 - 1.67 * e^{-K_{BL}} + 38.35 * \varepsilon_{ACA} \tag{7}$$

where $\varepsilon_{sol} = \|\mathbf{b} - \mathbf{A}\,\mathbf{x}\|_2$ (see eq. 2)

Coefficient K_{BL} depends not only on γ and N_{ind} but also on model geometry and mesh partitioner. The equations shown above are only applicable to the theoretical example (see Section 5) and geometric bisection partitioning. Different geometries may lead to different forms of dependencies but the tendencies are the same:

(i) The less N_{ind} yields higher K_{BL};
(ii) The higher γ yields higher K_{BL};
(iii) The higher K_{BL} yields higher solution error;
(iv) The higher ε_{ACA} yields higher solution error.

In most cases ε_{sol} and ε_{ACA} are of the same order of magnitude. So, in order to guarantee the solution error it is feasible to choose ε_{ACA} so that it is one order less the solution error, i.e. if the solution error $\varepsilon_{sol} = 10^{-5}$ then $\varepsilon_{ACA} = 10^{-6}$.

Let us consider how K_{BL} and ε_{ACA} affect memory consumption for different values of N. In order to simplify equations two dependencies have been measured: for floating K_{BL} keeping the same $\varepsilon_{ACA} = 0.001$ (8) and for different ε_{ACA} keeping the same $K_{BL} \approx 2.0$ (9).

$$\ln M(N, K_{BL}) = 15.1 + 0.056N^2 - 1.18K_{BL} + 0.57K_{BL}^{1.5} - 0.7\,K_{BL}^2 \tag{8}$$

$$\ln M(N,\varepsilon_{ACA}) = 5.7 + 1.34 \ln N - 0.093 \ln \varepsilon_{ACA} + 1.12/\ln \varepsilon_{ACA} \quad (9)$$

The evaluation of functions $T_{CPU}(N, K_{BL})$ and $T_{CPU}(N, \varepsilon_{ACA})$ gives very similar dependencies as for memory consumption M (8), (9) and shows that the higher K_{BL} as well as higher ε_{ACA} reduce T_{CPU} and M. There is a contradiction: from one hand, higher K_{BL} as well as higher ε_{ACA} reduce memory consumption but from the other hand they yield higher solution error. The solution of this contradiction is a question of balancing between involved parameters. There are three possible cases:

1. The solution has to be obtained as fast as possible and its error is not critical. In this case CPU time has priority. ε_{ACA} has to be chosen as high as possible and K_{BL} has to be optimal. An optimal value of K_{BL} depends on N_{ind}. Figure 3 shows that optimal value of N_{ind} is not the smallest one and therefore function $T_{CPU}(N_{ind})$ has a minimum in N_{ind}^{opt} - optimal value of N_{ind} for current cluster tree. This happens because very small values of N_{ind} lead to huge K_{BL}. As a result H-matrix represents a very rough approximation of the source matrix A (2) and iterative techniques for the solution of LSE need more iterations to converge. This fact obviously has a connection with the condition number of LSE, which goes up as ε_{ACA} grows. There is no special rule on how to find optimal value of N_{ind} for any cluster tree but in most cases this value is less than $0.05\,N$, where N is a total number of degrees of freedom.

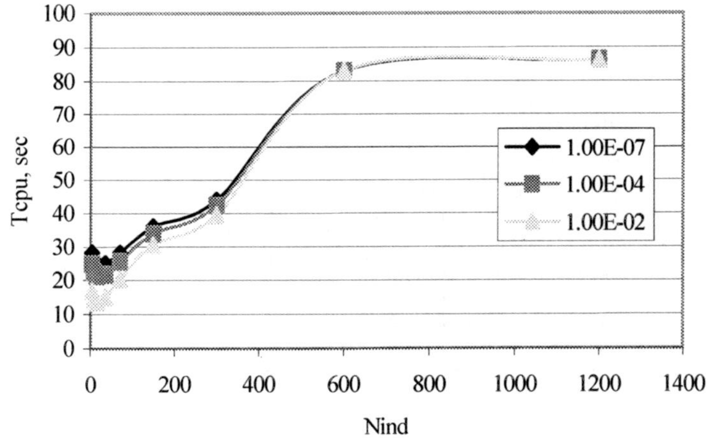

Figure 3: T_{CPU} vs. N_{ind} for different ACA-errors.

2. CPU time is not very critical but solution error has to be as small as possible. In this case K_{BL} and ε_{ACA} also have to be as small as possible. ε_{ACA} in this case has to be chosen one order less the needed solution error.

3. Memory capacity is limited and therefore memory consumption must be minimal. This case is very similar to the case number 1 because optimisation of CPU time is directly connected to minimisation of memory consumption. However, memory consumption does not have any tricks with N_{ind} and smaller N_{ind} leads to smaller matrix A (2) as lower N_{ind} generates more Rk-blocks which can be compressed up to chosen error ε_{ACA}.

5 Numerical examples

This section demonstrates efficiency of H-matrix technique in conjunction with ACA algorithm. The aim of these computations is to examine the performance of the ACA algorithm and to apply the results of the analysis made in Section 4. The ACA-based solver deals with Laplace equation with mixed boundary conditions: Dirichlet and Neumann. The boundary is meshed by linear triangular elements. The H-matrix storage approach and the ACA algorithm have been used as a black-box solver. As an input ACA-based solver takes matrix A (2) which has to be generated in advance by a BEM software, vector of coordinates for each DOF corresponding to each column and row in matrix A as well as the right-hand side vector of eq. (2). After ACA-approximation generalised minimal residual method (GMRES) is applied with the accuracy 10^{-6}. The solver is based on modified HLib library [7]. Geometrically, model is represented by a unitary cube (height, width and breadth = 1) with 5248 collocation nodes. A comparison of CPU time, memory capacity and error $\|\mathbf{b}-\mathbf{Ax}\|_2$ between the ACA-based solver with $N_{ind}=16$ and $\gamma=2.5$ and a Direct LU solver is presented in Table 1.

Table 1: Performance of ACA-based solver compared with Direct (LU) solver.

Solver	ε_{ACA}	T_{CPU}, sec	M, Mb	$\|\mathbf{b}-\mathbf{Ax}\|_2$
Direct (LU)	-	3870	220	0.0
ACA-based	10^{-6}	9.28	76	0.000006
	10^{-5}	7.58	60	0.000023
	10^{-4}	6.1	46	0.000174
	10^{-3}	4.76	34	0.002076
	10^{-2}	3.84	23	0.023374
	10^{-1}	3.43	14	0.698741

As a conclusion, ACA-solver gives reasonable speedup and saves memory needed for storing LSE. The ACA-solution with $\varepsilon_{ACA} = 10^{-3}$ still has good precision and compared to the direct solution it needs 6 times less memory and can be obtained much faster. Figure 4 shows the comparison between the result x_i obtained with ACA-based and Direct LU solvers, in the horizontal and vertical axes, respectively.

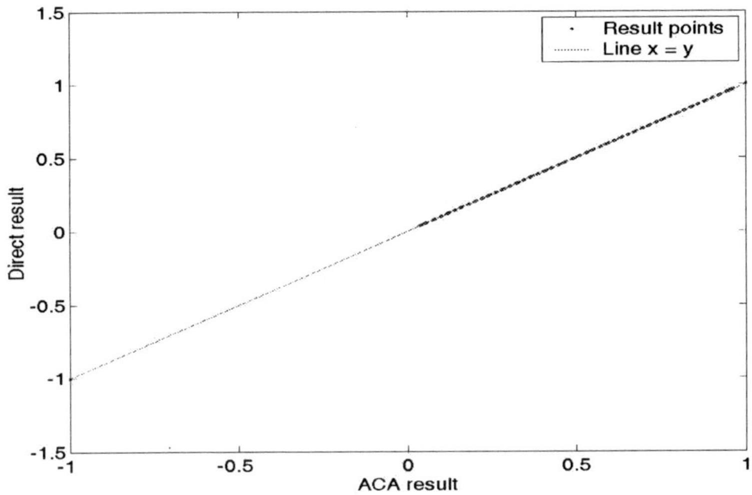

Figure 4: ACA- vs. Direct-solution comparison.

6 Conclusions

As a result of this work the main parameters that control the performance of an ACA-based solver for BEM have been explained and experimental dependencies have been established and cast into simple analytical expressions. A set of three possible cases of restrictions in computational resources have been established and on its basis appropriate recommendations regarding values of the parameters have been proposed. At the end, a practical toy example demonstrates the numerical properties of ACA-based solver for LSE and proves its efficiency.

Acknowledgements

The first author would like to thank the Wessex Institute of Technology for the support by scholarship and to express his gratitude to the Max-Planck Institute for their HLib library, which has been used in this work.

References

[1] Stefan Kurz, Oliver Rain and Sergej Rjasanow. *The Adaptive Cross-Approximation Technique for the 3-D Boundary-Element Method.* IEEE Transaction on Magnetics, Vol. 38, No. 2, March 2002

[2] Stefen Borm, Lars Grasedyck and Wolfgang Hackbusch. *Hierarchical Matrices*, April 2005

[3] M. Bebendorf and R. Grzibovski. *Accelerating Galerkin BEM for Linear Elasticity using Adaptive Cross Approximation*, May 2006

[4] Wolfgang Hackbusch, Lars Grasedyck and Steffen Borm. *An Introduction to Hierarchical matrices.* Proceedings of EQUADIFF 10, Prague, August 2001

[5] M. Bebendorf and S. Rjasanov. *Adaptive Low-Rank Approximation of Collocation Matrices.* Computing 70, 1-24, Springer-Verlag 2003

[6] Teresco James D, Karen D Devine, Joseph E Flaherty. *Partitioning and Dynamic Load Balancing for the Numerical Solution of Partial Differential Equations.* Numerical Solution of Partial Differential Equations on Parallel Computers, August 2005

[7] *Hlib library v.1.3.* Max Planck Institute for Mathematics in the Sciences. Inselstrasse 22-26, 04103 Leipzig, Germany

Author Index